EPA420-S-06-002
March 2006

Diesel Retrofit Technology

An Analysis of the Cost-Effectiveness of Reducing Particulate Matter Emissions from Heavy-Duty Diesel Engines Through Retrofits

Assessment & Standards Division and
Compliance & Innovative Strategies Division
Office of Transportation and Air Quality
U.S. Environmental Protection Agency

Executive Summary

The Environmental Protection Agency's (EPA) National Clean Diesel Campaign (NCDC) is a comprehensive initiative to reduce pollution from diesel engines throughout the country, including vehicles on highways, city streets, construction sites, and ports. The NCDC comprises both regulatory programs to address new engines and voluntary programs to address the millions of diesel engines already in use. On the regulatory side, EPA is successfully implementing emissions standards for engines in the 2007 Heavy-Duty Highway Engine Rule and the Tier 4 Nonroad Rule and developing new emission requirements for locomotives and marine diesel engines, including large commercial marine engines. On the voluntary side, EPA is addressing engines that are already in use by promoting a variety of innovative emission reduction strategies such as retrofitting, repairing, replacing and repowering engines; reducing idling; and switching to cleaner fuels. The voluntary programs are accomplished in partnership with state and local governments, environmental groups and industry.

The emissions standards for new engines will reduce both highway and nonroad engine emissions by roughly 90%. However, these emission reductions occur over a long period of time as new engines are phased into the fleet. Retrofitting diesel engines currently in use will allow significant and immediate emission reductions from diesel engines that would not otherwise be addressed.

The purpose of this technical analysis is to evaluate the cost effectiveness of retrofitting existing heavy-duty diesel engines to reduce particulate matter (PM). (The cost effectiveness of the regulatory measures EPA has implemented is addressed the rulemakings.) Analysts in EPA's Office of Transportation and Air Quality (OTAQ) evaluated the costs and emissions benefits of retrofitting school buses, freight trucks, and bulldozers with diesel oxidation catalysts (DOCs) and catalyzed diesel particulate filters (CDPFs), two of the most common PM emissions reduction technologies for diesel engines.

For highway vehicles (e.g. school buses and trucks), EPA considered two overarching methods to estimate the cost effectiveness of diesel retrofit technology. The first involved using only the current mobile source emission factors and inventories in EPA's approved MOBILE6.2 model. The second involved using more recent data that OTAQ has collected to use in the future development of EPA's next highway emissions model, MOVES (Motor Vehicle Emissions Simulator). EPA chose the second option for this technical paper in order to better reflect more recent information on highway vehicles.

EPA obtained the more recent highway vehicle data from states, fleet owners, and technology and engine manufacturers covering factors such as annual vehicle miles traveled, vehicle useful life, engine emission rates, retrofit technology effectiveness, and technology costs. For example, this paper assumes heavy-duty diesel PM emissions are approximately 2.3 times higher than predicted in MOBILE6.2 based on results from recent chassis dynamometer testing from the California Air Resources Board, the Coordinated Research Council, EPA's efforts to update the MOBILE model, and other sources. EPA will eventually use the more recent highway vehicle data to modify the MOBILE6 model as part of a comprehensive effort to create the next generation mobile model, MOVES. It is important to note, however, that states and local governments are still using MOBILE6.2 to estimate highway vehicle emissions for State Implementation Plans (SIPs) and transportation conformity purposes.

For nonroad engines (e.g. 250 hp bulldozers), EPA relied primarily on data from the NONROAD2004 model to determine the cost-effectiveness of DOCs. EPA also consulted additional data sources where appropriate.

EPA calculated that the cost effectiveness for both school bus diesel oxidation catalyst (DOC) and catalyzed diesel particulate filter (CDPF) retrofits ranged from $12,000 to $50,500 per ton of PM reduced. The same type of retrofits for Class 6&7 heavy-duty highway trucks (commonly found on highways and city streets) ranged from $27,600 to $69,900 per ton of PM reduced. The same type of retrofits of larger Class 8b trucks (commonly used to transport freight long distances) ranged from $11,100 to $44,100 per ton of PM reduced. Finally, DOC retrofits for 250 hp bulldozers ranged from $18,100 to $49,700 per ton of PM reduced.

The results can be compared to similar estimates for other EPA programs targeted at reducing diesel particulate matter. For example, EPA estimates that the cost-effectiveness of the Urban Bus Retrofit and Rebuild program is $31,500/ton of PM reduced, the 2007 Heavy-Duty diesel emission standards is $14,200/ton, and the Nonroad Tier 4 emission standards is $11,200/ton.

The findings from this study indicate that retrofits can be a cost effective way to reduce air pollution.

Table of Contents

I. INTRODUCTION

I.A. NATIONAL CLEAN DIESEL CAMPAIGN

The Environmental Protection Agency's (EPA's) National Clean Diesel Campaign (NCDC) is a comprehensive initiative to reduce pollution from diesel engines. EPA's Office of Transportation and Air Quality (OTAQ) manages the NCDC, which comprises both regulatory programs to address new engines and voluntary programs to address the millions of diesel engines already in use.

Particulate matter (PM), one of the primary pollutants from diesel exhaust, is associated with many different types of respiratory and cardiovascular effects, and premature mortality. EPA has determined that it is a likely human carcinogen. Fine particles (smaller than 2.5 micrometers), in particular, are a significant health risk as they can pass through the nose and throat and cause lung damage. People with existing heart or lung disease, asthma, or other respiratory problems are most sensitive to the health effects of fine particles as are children and the elderly. Children are more susceptible to air pollution than healthy adults because their respiratory systems are still developing and they have a faster breathing rate. EPA expects reductions in air pollution from diesel engines to lower the incidence of these health effects, as well as contribute to reductions in regional haze in our national parks and cities, lost work days and reduced worker productivity, and other environmental and ecological impacts.

New regulations from EPA require stringent pollution controls on new highway and nonroad diesel engines, including engines operating in the freight, transit, construction, agriculture, and mining sectors. The new regulations will also slash sulfur content in diesel fuel by 97 percent. By combining tough exhaust standards with cleaner fuel requirements, these rules will cut emission levels from new engines by over 90 percent. The new lower sulfur diesel fuel will result in reduced PM emissions as soon as the fuel is introduced into the market. New engines sold in the US after 2007 for highway use (and after 2010 for nonroad use) must meet the more stringent standards, but the effect of these cleaner engines will be achieved over time as the existing fleet is gradually replaced. The

benefits of these new rules will not be fully realized until the 2030 time frame. As a result EPA is promoting a suite of voluntary programs to address the emissions from the existing fleet of diesel vehicles.

The NCDC voluntary programs are designed to address existing diesel vehicles and equipment through emission reduction strategies that can provide immediate air quality and health benefits. The voluntary programs focus on vehicles and equipment in the school bus, construction, port, freight and agricultural sectors. The voluntary programs work with partners in state and local government, industry, and environmental organizations to promote a wide range of measures to reduce diesel emissions including retrofitting vehicles with new or improved emission control equipment, upgrading engines, replacing older engines with newer/cleaner engines, reduced idling, and using cleaner fuels.

I.B. STUDY OBJECTIVE & METHODS

Stakeholders - including states that are developing their plans to achieve the National Ambient Air Quality Standards for fine particles - are searching for cost effective ways to reduce emissions from existing diesel engines in order to improve air quality and protect public health. The purpose of this study is to estimate the cost effectiveness of retrofit strategies.

We chose to evaluate retrofit strategies for four types of vehicles:

   1) school buses
   2) combined class 6&7 trucks
   3) class 8b trucks, and
   4) 250 horsepower (hp) bulldozers

Truck classes are based on the Gross Vehicle Weight Rating. Class 6 trucks are 19,501 - 26,000 lbs and Class 7 trucks are 26,001 - 33,000 lbs. Class 6&7 trucks are commonly found on highways and on city streets. Class 8b vehicles are greater than 60,000 lbs and are commonly used to transport freight long distances. 250 hp bulldozers are technically called Diesel Crawler Tractors or Crawler Dozers in our NONROAD emissions inventory model, and are prevalent on construction sites around the country.

1

EPA chose these vehicle types for three reasons. First, we wanted to evaluate retrofits for both highway (e.g. school buses and trucks) and nonroad vehicles (e.g. bulldozers). Second, we had the best data for these types of vehicles due, in large part, to our experience with retrofit projects on-the-ground. Finally, these vehicles exist in large numbers across the country, so we believed that cost-effectiveness analysis for these vehicles would be relevant to a wide audience.

We decided to evaluate the two most common diesel retrofit technologies, diesel oxidation catalysts (DOCs) and catalyzed diesel particulate filters (CDPFs), for all vehicle types, except for 250 hp bulldozers, for which we only analyzed DOC retrofits since CDPFs are not currently compatible with many bulldozers.

For highway vehicles (e.g. school buses and Class 6-8b trucks), EPA considered two overarching methods to estimate the cost effectiveness of diesel retrofit technology. The first involved using only the current mobile source emission factors and inventories in EPA's approved MOBILE 6 model (version 6.2), OTAQ's emission factor model for predicting gram per mile emissions from cars, trucks, and motorcycles under various conditions. The second involved using more recent data that OTAQ has collected to use in the future development of EPA's next highway emissions model, MOVES. EPA chose the second option for this technical paper in order to better reflect more recent information on highway vehicles.

EPA obtained the more recent highway vehicle data from states, fleet owners, and technology and engine manufacturers which impacted factors such as annual vehicle miles traveled, vehicle useful life, engine emission rates, retrofit technology effectiveness, and technology costs. For example, this paper assumes heavy-duty diesel PM emissions are approximately 2.3 times higher than projected in MOBILE6.2 based on results from recent chassis dynamometer testing from the California Air Resources Board, the Coordinated Research Council, EPA's efforts to update the MOBILE model, and others. EPA will eventually use the more recent highway vehicle data to modify the MOBILE6.2 model as part of a comprehensive effort to create the next generation mobile model MOVES (Motor Vehicle

Emissions Simulator). It is important to note, however, that states and local governments are still using MOBILE 6.2 to estimate highway vehicle emissions for State Implementation Plans (SIPs) and transportation conformity purposes.

For nonroad engines (e.g. 250 hp bulldozers), EPA relied primarily on data from the NONROAD2004 model to determine the cost-effectiveness of DOCs. EPA also consulted additional data sources where appropriate.

For both highway and nonroad vehicles, we analyzed annual vehicle miles traveled, vehicle useful life, engine emission rates, retrofit technology effectiveness, and technology costs to calculate the cost-effectiveness of retrofit strategies, in terms of $ per ton of PM reduced. It is important to note that, in many cases, heavy-duty diesel retrofit strategies provide other emission benefits such as reductions in hydrocarbons and carbon monoxide. This study only evaluates the cost-effectiveness of reducing PM from diesel retrofits.

The following section will detail our methods for calculating the cost-effectiveness of PM reductions from retrofits including factors such as vehicle activity, survival rates, emissions factors, costs of technologies, and emissions reductions from retrofit technologies. In Section 3 we will present our results and in Section 4 we will provide summary remarks about the relative cost-effectiveness of diesel retrofits. As mentioned previously EPA calculates these cost effectiveness figures based on more recent information for highway vehicles obtained from various sources. If EPA chose not to use this more recent information and instead relied exclusively on the MOBILE model for these calculations, the cost effectiveness could range from approximately $14K to $160K per ton.

II. RETROFIT EFFECTIVENESS FACTORS

In order to estimate the relative cost effectiveness of various PM retrofit strategies, it is necessary to estimate a number of factors, including:

> -vehicle activity
> -vehicle survival rates

-emissions rates of vehicles
-effectiveness of DOCs and CDPFs
-costs of retrofits

The following sections 2.A - 2.G outline our methodologies for estimating each of these factors.

II.A VEHICLE ACTIVITY

One of the first steps in estimating emission reductions from retrofit strategies is to develop an estimate of annual vehicle activity. This requires identifying nominal values for vehicle miles traveled for representative vehicle samples, in the case of highway vehicles (e.g. trucks and school buses), and operating hours and load for nonroad vehicles (e.g. bulldozers). This information can then be used to estimate annual vehicle emissions and emission reductions from retrofits.

II.A.1 School Bus Activity Analysis
The default MOBILE 6.2 Vehicle Miles Traveled (VMT) for school buses is 9,939 miles per year [1]. Anecdotal reports suggest that average school bus VMT has increased over time. This increase is attributed to suburban growth around many communities at a time when budget-strapped school districts cannot afford to expand their school bus fleets.

As a test of this anecdotal information, we reviewed detailed school bus fleet data that school districts submitted to EPA in response to a request for applications for Clean School Bus USA grant funding over the summer of 2003. The Clean School Bus USA demonstration

grants program attracted 120 applications from diverse programs around the country seeking to retrofit or replace aging school bus fleets. Of these, 72 applications contained data that were relevant to this exercise and that were in a format that could be analyzed. Most of the applications provided actual fleet VMT data from the 2002-2003 school year, with the others submitting data from 2001-2002. The data represent several hundred school districts, and more than 34,000 school buses from 31 states plus Puerto Rico.

We analyzed the average school bus activity only for diesel school buses.

We took the average per-bus VMT for each fleet directly from the application, or, if not provided, calculated it by dividing the annual fleet mileage by the number of buses in the fleet. If a fleet's total mileage included diesel and non-diesel school buses, we weighted the annual mileage by the technology ratio to reflect only the diesel portion of the fleet. If applicants provided only total mileage and age for each bus in the fleet, we calculated an average VMT for each bus by dividing mileage by age, and then created a fleet average by averaging the VMTs from individual buses.

To determine a representative VMT mix across the population of buses considered for this analysis, we calculated a fleet fraction for each school bus fleet by dividing the number of buses in that fleet by the total population. We then multiplied that fraction of the population by the fleet's average VMT to create a weighted fleet fraction of the fleet's VMT. We determined the average VMT for the total population by adding the weighted fleet fractions.

The method described above yielded an average annual VMT of 13,248 miles per bus. This is an increase of approximately 3,309 miles per year from the default value currently in the MOBILE6.2 model and is used to represent average VMT for school buses independent of vehicle age. That is, school bus VMT is estimated to be the same in the first year and all subsequent years of the vehicles life. Although this represents a simplification of real-world practices, we believe that, given the fixed routes defined for many school buses, this is a reasonable assumption. However, there are no independent data available to test the assumption from this analysis.

It is important to note that the annual school bus VMT used in our analysis represents a relatively large increase over the school bus VMT estimate from the Vehicle Inventory and Use Survey (VIUS) study (the basis for the values used in the development of MOBILE6.2).

II.A.2. Truck Activity Analysis

We used an estimate of annual VMT from MOBILE 6.2 for Class 6-8b trucks which declines with vehicle age. This estimate can be found in EPA report, Fleet Characterization Data for MOBILE6: Development and Use of Age Distributions, Average Annual Mileage Accumulation Rates and Projected Vehicle Counts for Use in MOBILE 6.2 (see Table 1 Annual Mileage Accumulation).[2]

II.A.3 Nonroad Activity Analysis

Our methodology for estimating emission reductions from nonroad equipment is similar to that for highway vehicles in that we first needed to estimate annual and lifetime activity (use patterns). We estimated this activity based on data from the technical documentation for the NONROAD inventory emissions mode (see www.epa.gov/otaq/nonrdmdl.htm for a description of the NONROAD model). Nonroad engine activity is expressed in terms of hours of operation (annual and lifetime) and load factor (average engine operating power as a percentage of rated engine power). The estimate for annual hours of operation for a 250 hp bulldozer is 936 hours per year. The estimate for

the typical load factor for a 250 hp bulldozer is 0.59 (average cycle power/rated power).[3]

II.B. VEHICLE SURVIVAL RATE

The scrappage rate describes the fraction of vehicles (relative to the total number originally sold) that are no longer in the fleet from one year to the next. This factor reflects vehicle loss through accidents, deterioration, and export. From a retrofit perspective, scrappage is a necessary component of cost effectiveness analysis because it dictates how long older vehicles will stay on the road, and hence the potential benefit which will accrue from a retrofit at a certain point in time.

II.B.1. Highway Scrappage Analysis

An analysis of scrappage rates for selected model years of Heavy-Duty vehicles is published by Oak Ridge National Laboratory in the Transportation Energy Data Book (TEDB), based on registration data and a scrappage model developed by Greenspan and Cohen[4]. The latest model year for which TEDB published data on scrappage rates is 1990, but we did not use these data for our analysis because they seemed unrealistically high - for example, they projected a 45 percent survival rate for 30 year-old trucks. While limited data exist to confirm this judgement, a snapshot of 5-year survival rates can be derived from the VIUS for 1992 and 1997 for comparison. According to VIUS, the average survival rate for model years 1988-1991 between the 1992 and 1997 surveys was 88 percent. The comparable survival rate for 1990 model year Heavy-Duty vehicles from the TEDB was 96 percent, while the rate for 1980 model year trucks was 91 percent. Based on this analysis, we determined that 1980 model year survival rates are more in line with available data, and these rates are used in the analysis instead of the 1990 rates. The resulting median life estimate (the age at which 50% of vehicles have been scrapped) is 18.5 years. This contrasts with a median life estimated in the MOBILE6.2 emission model of approximately 12 years. The difference between the two data values indicates that there is some degree of uncertainty regarding survival rates. Survival Rates are shown in Table 2.

II.B.2 Nonroad Scrappage Analysis

Like the MOBILE model, the NONROAD model has intrinsic scrappage rates built into the model. These rates are used to project the distribution of nonroad equipment in a population by age. We chose to use a simplified nonroad scrappage rate estimate for this analysis. We use the resulting median life estimate for nonroad equipment in the NONROAD model. This number is the number of hours of rated engine operation that the median example of nonroad diesel engine is expected to operate. Dividing that number by the load factor (discussed previously) converts the median life from hours of operation at rated power to hours of operation at typical operating power levels (i.e., it converts it to actual hours of operation). The median life for a 250 hp diesel engine from the NONROAD model is 4,667 hours at rated power. Dividing this number by the typical load factor found previously (4,667 hours rated / 0.59) returns a median life at typical operating conditions of 7,910 hours. Given annual operating hours of 936 hours, the expected lifetime for the median 250hp nonroad bulldozer can be found as 8.5 years. While this represents the expected median operating life, it should be recognized that significant variation about this median can be expected in practice with many pieces of nonroad equipment being used for periods well in excess of 8.5 years.

II.C. EMISSION RATES

MOBILE6.2 is the current, approved highway emission factor model used by States and local governments for State Implementation Plans (SIPs) and planning purposes. When the analysis portion of MOBILE6.2 was completed in 1998, there were little heavy-duty chassis dynamometer data available, therefore the emission rates in MOBILE6.2 are based on engine dynamometer test data from engine certification tests. With chassis dynamometer testing, the engine remains in its vehicle (chassis) and the vehicle's tires drive rollers that produce a load in the dynamometer. This produces a more realistic test of the engine in this application than an engine dynamometer test where the engine is removed from the vehicle and connected directly to a dynamometer. The analysis for highway vehicles in this report uses more recent data, which EPA is planning to incorporate into EPA's next highway emissions model, MOVES, when it is developed.

II.C.1. Highway Emission Rate Analysis

OTAQ completed a large data collection effort in 2002 and 2003 on Heavy-Duty Highway vehicle emissions rates. There have been several Heavy-Duty chassis dynamometer test programs completed in recent years which we obtained from EPA's Mobile Source Observation Database (MSOD) for this study.

While there is a significant number of Heavy-Duty diesel Class 8 chassis dynamometer tests in the MSOD, there is a lack of school bus chassis dynamometer emissions tests and limited tests for Heavy-Duty diesel Class 6&7 vehicles (which are similar in size to many school buses). For this analysis we developed a ratio metric correlation for Heavy-Duty diesel Class 8 emissions from chassis dynamometer tests to engine dynamometer test data represented in MOBILE6.2, and applied this ratio to Heavy-Duty diesel Class 6-8 MOBILE6.2 emissions to estimate emission rates for those vehicle classes. We made this estimate by applying the correlation described below to the current emission factors for Class 6-8 vehicles from the MOBILE 6.2 model to create new emission factors that should more closely match actual emission rates in use. This approach inherently assumes the ratio relevant to Class 8 vehicles applies to Class 6 and 7. While we believe this approach is reasonable, we do not have independent data to validate this assumption.

The available chassis data set consisted of 39 vehicles tested on 20 different cycles for a total of 315 tests. The bulk of the data used federal diesel grade 2 or California diesel fuels. We eliminated other fuel types from the analysis. See Table 3 for a description of the data set by cycles, model years, and fuel type. Table 4 shows a brief description of the cycles. More detailed plots of the test cycles are available upon request by contacting MOBILE@epa.gov.

We performed an analysis to determine if there was a trend by model year for the chassis to engine test results comparison. We did not find any obvious trend, and a weighted regression did not prove to be statistically significant. We compiled average PM emission rates for each model year on each of the two fuels used from the chassis data set. We computed the ratio for each model year of the chassis data set to the MOBILE6.2 emission rate, for each fuel type. When we took an average - including both fuels - weighted by sample size, over each model year, which resulted in a ratio of 2.3 for chassis dynamometer emissions to engine dynamometer emissions. This ratio reflects the effects of real world driving, in-use deterioration, and in-use fuels. See Table 5 for the emission rates and ratios by model year for the data set described above versus MOBILE6.2. In order to make a conclusion on impacts of these in-use factors on actual PM inventories from highway vehicles, we would have to conduct a detailed analysis isolating the effects of cycles, fuel, and deterioration to estimate the effects over the range of model years and vehicle classes available in MOBILE6.2. We felt, however, that the use of the ratio approach (described above) to determine emission rates is appropriate for this cost effectiveness analysis. EPA will consider this information as it develops its future emissions models that are to be used in the future for official SIP or conformity purposes.

It is important to note that the ratio we use in this analysis also does not appear in the newly released Retrofit benefits calculation module of the National Mobile Inventory Model (NMIM) which is based on MOBILE 6.2 emission rates. The benefits in the NMIM model come from a percentage tagged to each different technology used for retrofit. EPA would need to analyze a substantial amount of additional data to update the emission rates used in MOBILE 6.2.

For school buses we made one additional change. The PM emission factors in MOBILE6.2 are based on inputs from an earlier mobile source model called PART5. The PART5 model did not specifically identify emission factors for school buses, but it does contain an estimate for "buses", a category which would include urban transit buses, coach buses, and school buses. MOBILE6.2 does have an emission factor for school buses but that factor is simply a carryover from the bus emission factor in PART5. The bus emission factor in PART5, and hence, the school bus emission factor in MOBILE6.2 is based primarily on data and emission standards for urban buses. This causes two problems for school buses. First, older school buses have emission factors that are likely to overestimate PM emissions by approximately 50 percent due to the use of the higher emission conversion factor for urban buses. Second, emission factors attributed to new school buses are too low due to the use of the lower urban bus emission standard to project future emission rates even though school buses do not need to meet the lower standard (the current PM standard for school buses is twice that of urban buses, 0.1 g/bhp-hr versus 0.05 g/bhp-hr). Therefore, in this analysis we have chosen to model the school bus emission factor the same as combined Class 6&7 trucks. Class 6&7 trucks are most similar to diesel school buses using the same engines meeting the same emission standards.

DOC on construction equipment

II.C.2 Nonroad Emission Rate Analysis
The NONROAD engine model uses emission rates for nonroad diesel engines based on the emission standards, historic engine certification data, and projections of in-use deterioration of emissions over the lifetime of the equipment. Additionally, the nonroad model includes a factor to correct for observed differences in emissions production between in-use operating cycles and

the steady-state emissions test results. The projected in-use emissions rates are therefore the product of the expected new certification emissions level, the ratio of transient emission rates to steady-state emission rates, and projected deterioration rates over time (i.e., as the equipment ages EPA projects emissions will increase). The result of this methodology is that new (beginning of life) nonroad equipment is estimated to have a lower emission rate than the same equipment would after a period of operation. In order to simplify the analysis in this paper, we have combined the adjustment for transient emissions and deterioration into a single static number of 1.5 (i.e., a 50% increase in emissions over the certification levels) which roughly approximates the combined factors for a bulldozer in the nonroad model. This approach may undercount the emissions from a typical piece of nonroad equipment when compared to the NONROAD model where the transient adjustment factor ranges from 1.23 to 1.97 and the deterioration factor varies from 0 at 0 hours to 0.473 at full useful life.[5] Hence, the NONROAD model adjustment would range from 1.2 to 2.9 (1.0 X 1.23 to 1.473 X 1.97) over the range of engines and through the equipment life. We believe the use of a simplified single value of 1.5 is appropriate for this analysis since our goal is to estimate a nominal ratio of emission reductions and cost. EPA has developed a retrofit modeling function within NMIM that fully incorporates the features of the NONROAD model and will allow states and local authorities to more accurately estimate the potential for emission reductions through retrofits.

II.D. COMPARISON TO EXISTING HIGHWAY EMISSIONS INVENTORY MODEL

As mentioned above, MOBILE6.2 is EPA's official emissions factor model for highway Heavy-Duty engines and vehicles. A complete description of MOBILE 6.2 can be found on EPA's web site at www.epa.gov/otaq/mobile.htm.

In the previous sections (II.A - II.C), we analyzed a number of factors to estimate emissions from highway Heavy-Duty engines (Class 6&7, Class 8b, and school buses) and newly developed estimates based on the most recent data, where appropriate. These factors are annual VMT for

school buses, scrappage rates for Heavy-Duty vehicles, and a new engine-to-chassis conversion factor. Table 6 is a comparison of the estimates we developed for this paper and the emissions inventory values in MOBILE6.2.

Table 6. Comparison of MOBILE6.2 Values to Retrofit Cost Effectiveness Study Values

Factor	MOBILE6.2 value	Retrofit Analysis value
School Bus VMT	9,939	13,248
Scrappage	12 year median life	18.5 year median life
Engine Based PM Emission Factor[a]	1991MY 1.948 2000MY 0.163	1991MY 0.518 2000MY 0.158
Engine-to-Chassis Conversion	1 (no factor used)	2.3

[a] The engine based PM emission factors change for each model year, two example years are shown here.

When considered as a whole, the estimated values of lifetime emissions that we developed for this analysis (detailed in Sections II.A - II.C) are approximately 3 times greater for newer school buses and 2.3 times greater for Class 6-8 vehicles than the values used in MOBILE6.2. This general characteristic does not hold for model year 1993 and older school buses due to the much higher emission factors in MOBILE6.2 for these vehicles when compared to this analysis. The results for 1993 and older school buses are approximately 20 percent lower in this analysis when compared to estimates from MOBILE6.2. Section 3.Ci explains our rational for using alternative emission factors for school buses in this analysis.

II.E. EFFECTIVENESS OF RETROFIT TECHNOLOGIES

II.E.1. Background on Retrofit Technology Verification
The NCDC voluntary programs encourage air quality agencies and owners of fleets of diesel powered vehicles and equipment to implement

clean diesel strategies such as installing new or enhanced emission control technology and using cleaner fuels. To help these organizations make informed decisions regarding which retrofit technologies are appropriate for their fleets and what emission reductions can be expected, EPA created the Retrofit Technology Verification Process. This process evaluates the emission reduction performance of retrofit technologies, including their durability, and identifies engine operating criteria and conditions that must exist for these technologies to achieve those reductions.

Under this program, companies can apply for EPA verification of the effectiveness of their emission control technology. The verification protocol requires the same tests as defined by the Code of Federal Regulations (CFR) for new engine family certification before sale in the U.S. The protocol tests the stand-alone engine, and then the engine with the emission control technology. Both new and aged devices must be tested. The emission reduction percentage that EPA verifies will reflect the performance of the new and used devices. Once a technology is verified, the company receives an official EPA verification letter, and the technology is listed on EPA's web site as a verified technology. There is no restriction on who may apply for verification. To date, EPA has verified nearly 20 technologies from different emission control technology companies.

The measures that EPA verifies can be very general - for example, an emission control technology company may receive verification for a diesel oxidation catalyst (DOC) technology that can reduce particulate matter from any uncontrolled or Tier 1 nonroad diesel engine by 20 percent - or the verification can be specific to an engine model made over specific model years.

While retrofit technologies are the most common clean diesel strategy verified by EPA, there is a wide range of measures that can reduce diesel emissions. For example, the replacement of older engines or vehicles may be more beneficial in many cases than using retrofit technologies. If a fleet manager is concerned that exhaust emissions from their vehicles may overwhelm current retrofit technologies, or they are interested in having more up-to-date safety

equipment, the fleet manager may prefer to replace older vehicles with newer models rather than retrofit their existing vehicles.

II.E.2 Analysis
We took the retrofit technology applications and emissions reduction information in this cost effectiveness study directly from EPA's List of Verified Technologies. We are focusing only on emission reduction figures for DOCs and CDPFs verified for Class 6&7 Heavy-Duty engines.

The estimated reduction in PM:
1) from adding a DOC to a highway engine is 20%
2) from adding a DOC to a nonroad engine and changing to highway fuel ($\leq$ 500 ppm S) is 20%
3) from adding a CDPF to a highway engine and changing to ultra low sulfur diesel (ULSD) fuel from regular highway diesel fuel is 90%

One requirement of the verification process is that applicants must test their systems after they have been installed for a period of time. The manufacturer must begin in-use testing after they have sold a certain number of units of the verified system. EPA must approve the manufacturer's sampling plan to gather units to be tested. The manufacturer must test units aged in the field to a minimum fraction of the designated durability testing period in two different phases. Manufacturers are given wide latitude in the type of emissions testing equipment they use, although test cycles are well defined. The manufacturer must test at least four units in each phase. Individual failures lead to additional testing or possible removal from the Verified Technology List. This part of the verification process is still in its infancy and, as such, EPA has not yet received any in-use test results from retrofit technology manufacturers. Once EPA receives these additional in-use test results, EPA will examine them and use them to help quantify real world retrofit benefits.

The reduction of other criteria air pollutants by aftertreatment devices should also be recognized. A DOC or CDPF may reduce hydrocarbon and carbon monoxide emissions on the order of 20 to 90 percent.

II.F COSTS

II.F.1. Background

Several sources of information are available on the current price of retrofit technologies. These include a December 2000 survey by the Manufacturers of Emission Controls Associations (MECA) and current price information for grant recipients in EPA's Clean School Bus Program.[6] Those sources give ranges for CDPF prices of $3,000 to $7,500 depending on size, expected product sales volumes, and configuration (i.e., in-line or muffler replacement). Similarly, these sources suggest DOCs will range in price from $425 to $1,750 depending on size, sales volume and configuration. While we believe these ranges are reflective of current prices for PM retrofit technologies applied to Class 6&7 trucks and school buses, we also believe that future retrofit costs are likely to drop substantially as a result of the Heavy-Duty 2007 emission standards and the Nonroad Tier 4 emission standards.

II.F.2. Highway Cost Analysis

For this report, EPA has conducted a review of available information on the cost of PM retrofit technologies and has made a new projection for the future cost of PM retrofit technologies in 2007. Beginning in 2007, all new Heavy-Duty diesel engines are required to meet a PM emissions standard of 0.01 g/bhp-hr. EPA projected in the 2007 rule-making that this emission standard would be met through the use of CDPFs. Our recent Highway Diesel Progress Review Report #2 confirms that all Heavy-Duty diesel engine manufacturers are planning to comply with these regulations through the use of CDPF technologies.[7] This means that, beginning in the 2007 model year, the market - and therefore production volumes - for CDPFs will increase from a few thousand units a year in the United States to almost a million units a year. At the same time, there is increasing demand in Japan and Europe spurred by retrofit programs and new emission standards. In the aggregate, CDPF production volumes are expected to increase by almost two orders of magnitude (i.e., from tens of thousands annually to more than a million annually). In manufacturing, substantial cost savings can typically be found with increasing production volumes, especially when those production volumes change by orders of

magnitude. Therefore, we expect the cost for CDPFs to decrease significantly after 2007 compared to today's costs. For this analysis of the cost effectiveness of future retrofit programs, we feel it is appropriate to make an estimate of the future cost of retrofit technologies rather than relying on today's costs.

EPA has recently made an estimate of the production cost for CDPFs in the Nonroad Tier 4 rule-making.[8] The analysis in that rule-making was based on preliminary data available to EPA regarding the actual manufacturing costs for CDPF and DOC technologies. We decided to use that analysis as a basis for our projection of the future retrofit cost for Class 6&7 trucks and school buses. We have made a number of additions and modifications to the Tier 4 analysis to account for differences between high volume engine manufacturing and retrofit applications. Specifically, we have added additional costs to account for the instrumentation and testing necessary to qualify candidate retrofit vehicles for CDPFs and for the installation of the retrofit technologies. We have also accounted for additional canning and packaging costs specific to retrofit technologies and for the differences between projected manufacturing costs in 2007 (the period for this analysis) and the Tier 4 time frame (post 2010).

In order to ensure successful application of passive CDPF technologies, retrofit companies typically instrument a sample of candidate vehicles from a retrofit fleet to confirm that operating conditions and exhaust temperatures are appropriate for CDPF regeneration. Absent such testing, CDPFs can inadvertently be installed on vehicles for which passive regeneration is not assured, potentially leading to CDPF failure. We estimated the cost for this testing at approximately $2,000 dollars per twenty vehicles retrofitted (i.e., the cost for the testing is estimated as $2,000 dollars and the results from the test are assumed to be, on average, applicable to twenty vehicles within a fleet). Thus, we estimated an average cost of $100 per CDPF vehicle retrofit to account for the total cost of this testing.

The labor associated with installing a catalyst technology in a vehicle on a production line is quite small and not substantially different from the cost of installing an exhaust system.

Installing a catalyst may entail more labor if the catalyst weighs more than an exhaust system or needs additional fasteners, or connection points. The labor cost for installing retrofit technologies, however, can be a significant fraction of the overall cost. Installing a retrofit catalyst may entail removing a portion of the existing exhaust system, on-site fabrication or welding of connections to the exhaust system, and then remounting of the exhaust system. The facilities available for retrofit installation, typically vehicle service facilities, are also not designed as efficiently as vehicle assembly lines when it comes to installing a single component on vehicles. For these reasons, we felt that it was necessary to account for the additional installation cost (primarily labor) of retrofit technologies in this analysis. To accomplish this, we used data from the grant proposals provided to EPA under the Clean School Bus USA program. A number of the grant proposals included a cost for installation of the retrofit technologies. The average installation cost from these grant proposals was $193. We have used this average as an estimate for the installation of both CDPFs and DOCs for Class 6&7 trucks and school buses. Although, it might be reasonable to assume this cost will decrease in the future, we do not have adequate information to project the degree to which this average cost might change.

CDPF Installation

In addition to higher labor costs related to installing retrofit technologies (relative to the volume of vehicle production) we also expect there to be additional hardware costs associated with unique fastening and mounting systems for retrofit technologies. This reflects the fact that older vehicles were not designed to accommodate PM control after-treatment technologies. We have estimated the cost for these additional hardware components (additional fasteners and perhaps unique exhaust tubing and fittings) at $87 per vehicle for DOC retrofits and $300 per vehicle for CDPF retrofits. It may be possible in the more distant future that these components will reach a degree of commonality that will lead to lower costs, however, at this time we did not have enough information to estimate the degree to which these costs may change.

In order to apply the Nonroad Tier 4 Regulatory Impact Analysis estimate of CDPF and DOC costs for retrofit vehicles, we needed to address the difference in time horizons for the two future projections. This analysis is intended to project the cost for retrofit technologies in the year 2007, while the Nonroad Tier 4 analysis focused on technology cost in 2011 and beyond. Reflecting a start date beyond 2010, the Nonroad Tier 4 analysis incorporated a 20 percent learning curve effect into its estimate of future CDPF costs (no learning curve effect was applied for DOCs). For this analysis, we have removed that learning curve effect in order to correlate the estimate to an earlier time period, specifically 2007. The resulting cost for a CDPF (without the other costs noted previously) is $1,920 for a diesel engine of 8 liter engine displacement. The DOC cost (again without the additional costs listed previously) is $260 for Class 6&7 trucks and school buses.

Table 7 summarizes the total estimate we have made for the cost of PM retrofit technologies in the 2007 time frame. The table shows a projected cost for DOCs of $540 per Class 6&7 truck and school bus retrofitted and a projected cost for CDPFs of $2,500. These projections represent our best estimate of the nominal cost for retrofitting vehicles with diesel engines of 8 liter displacement. In practice, we would expect significant variability above and below these price estimates due to a wide range of other factors that we did not account for in this analysis (e.g., retrofit fleet size, profit margin differences, etc.). Nevertheless, we believe these estimates adequately reflect the nominal cost for future PM retrofit technologies.

The cost analysis described above is specific to engines with displacements of 8 liters applied to Class 6&7 trucks and school buses. In order to estimate the cost to retrofit larger Class 8b vehicles powered by engines with displacements typically between 11 and 16 liters, we have scaled this analysis by a ratio of 13:8 (i.e., we have increased the cost by 62 percent). This increase implicitly assumes that the retrofit cost is directly proportional to engine displacement and that 13 liters is a typical Class 8b engine displacement. Because many of the retrofit components are sized in direct proportion to engine displacement, we believe this approximation is robust. The resulting cost estimates for Class 8b retrofits are $880 per DOC retrofit and $4,100 per CDPF retrofit. As noted above, this estimate represents a nominal cost and a number of factors could result in costs that are lower or higher than those we have estimated.

II.F.3. Nonroad Cost Analysis

For our nonroad example application (250 hp bulldozer), we have taken a different approach to estimating the cost for future retrofit application. We have used a nominal average cost based on our current experience with nonroad retrofits. That typical cost is $800 per DOC retrofit on nonroad equipment. We have not made a future projection of reductions in this cost, because of the greater diversity and smaller retrofit fleet sizes typical of nonroad equipment. We expect nonroad retrofits to occur one piece of equipment at a time, even in relatively high volumes. We believe using today's nominal cost as a future cost estimate is very conservative, but given the uncertainty in the nonroad retrofit market we do not attempt to predict future cost reductions.

II.F.4. Highway and Nonroad Operating Costs

We do not account for operating costs related to the application of PM retrofit technologies in this analysis. Operating costs could include the differential cost for using 15 ppm sulfur fuel, fuel economy impacts related to increased exhaust backpressure, or changes to maintenance practices related to the use of retrofit technologies. We have not accounted for a 15 ppm sulfur fuel premium in this analysis because in 2007 (the time frame of this analysis) 15 ppm

sulfur highway diesel fuel will be the predominant diesel fuel used in highway applications. At the same time nonroad engines must switch to fuel with less than 500 ppm sulfur. We have not accounted for a change in fuel consumption related to the use of PM retrofits in this analysis because current data from existing retrofits show no significant difference in fuel economy for vehicles with and without PM retrofit technologies.[9] In practice, the impact of retrofit technologies on fuel consumption is strongly related to engine load and therefore varies significantly depending upon the vehicle application.

In the HD2007 rulemaking, we made estimates of the lifetime operating costs for maintenance related to cleaning accumulated oil ash from CDPFs. Those costs reflect a net present value calculation (in the year of sale) for a future maintenance cost that would occur after 150,000 miles of trap operation. We project, however, that only a limited number of retrofitted vehicles in Classes 6&7 will accumulate 150,000 miles of operation after the CDPF retrofit. In most cases, we project that vehicles will be scrapped prior to the time when this maintenance would be necessary. Therefore, while some vehicles will receive this maintenance (for example, vehicles with higher annual VMT than projected in this analysis), we have not accounted for these maintenance costs for school buses and Class 6&7 trucks in this analysis. For Class 8b trucks, which tend to accumulate many more miles, we have included the maintenance cost estimated in the HD2007 rulemaking of $208.

II.G. ESTIMATING LIFETIME EMISSION REDUCTIONS

II.G.1. Background

In order to compare the relative cost effectiveness (i.e., tons of emissions reduced per dollars spent) of PM retrofit programs to other PM emission control programs, it is necessary to estimate the lifetime emissions reduction we project will occur with PM retrofits. In concept, estimating the emission reductions is simple and can be viewed as the product of the lifetime vehicle miles traveled (VMT), the baseline emission rate for the vehicle (grams/mile) and the emission reduction potential of the retrofit technology (e.g., 90% for CDPFs). In practice, the estimation is more complicated since we must account for vehicle scrappage, variations in vehicle miles traveled as the vehicle ages, and the relative value of emission reductions realized in the current year versus a future time. Furthermore, estimates of the lifetime emission reductions for retrofit technologies must address the age of the vehicle when the retrofit is installed (i.e., retrofitting a one year old vehicle would be expected to result in a larger emission reduction compared to a ten year old vehicle). We have accounted for these factors in our analysis for the nominal case, but it should be recognized that factors such as annual vehicle miles traveled can vary significantly between different vehicles.

II.G.2. Highway Emission Reduction Analysis
Earlier in this report, we provided an estimate of the average annual VMT for school buses participating in the Clean School Bus USA program and for vehicles with Class 6-8 Heavy-Duty engines. These estimates for annual VMT reflect the mileage a vehicle may travel in each full 12 month period of its operating life. However, some vehicles will invariably be scrapped prior to reaching their total potential lifetime VMT. So while we estimate that a 20 year old school bus may have an average VMT of 13,248 miles per year in this analysis, we would only expect a small percentage of school buses to remain in operation after 20 years. In a previous section, we described the methodology used to estimate the fraction of vehicles that survive to a particular age based on historic registration data. Those data show, for example, that after 10 years 83% of trucks are expected to be in operation and conversely that approximately 17% will have been scrapped. Using this information, we can weight the annual VMT (and hence emission reductions) of a ten year old vehicle by the likelihood that the vehicle is still in use and generating those emissions or emission reductions. We accomplish this by multiplying the annual VMT of a ten year old vehicle and the survival fraction of ten year old vehicles. We make a similar calculation for every year of a nominal vehicle's life.

This approach allows us to estimate the emissions of a group of newly built vehicles, but it is somewhat problematic for retrofits of older vehicles. This is because the subset of retrofit

vehicles represent a surviving fraction from which the scrapped vehicles have already been removed, and for this analysis scrappage must be tracked for the retrofitted fleet according to when the retrofits were performed. For example, if we retrofitted a fleet of ten year old vehicles the scrappage rate for those vehicles in their first year of operation would be the age one scrappage rate of 0%, rather than the age ten scrappage rate of 17%. In order to account for the fact that retrofits of older vehicles begin with a subset of the survivors, we have created separate survival curves for retrofit vehicles of various ages from one year old to 28 years old at the time of the retrofit. We generated these survival curves by normalizing the survival fraction to 100% in the first year of operation, thus maintaining the general characteristics of the survival curve while reflecting the fact that retrofit vehicle groups are assured to have survived to the first year of their retrofit. Table 8 shows the survival fractions based on vehicle age and vehicle age at time of retrofit.

Using the information in Table 8 and the annual VMT estimates for school buses and Class 6-8 Heavy-Duty trucks, it is possible to make an estimate of weighted annual VMT for retrofit vehicles accounting for the survival fraction and the age of the vehicle at the time of retrofit. These estimates are presented in Tables 9, 10 and 11.

Based on the estimate of the nominal annual VMT for retrofit vehicles and weighting this VMT by the surviving fraction of a subset of vehicles retrofitted at a certain age, we can use this information to estimate the annual emission reductions for retrofit technologies as the product of the weighted annual VMT (Tables 9-11), the emission rate per mile, and the emission reduction (percent reduction) realized from the retrofit for each year of a retrofit vehicle's life. Tables 12 - 17 show the results of these calculations. Tables 12 - 17 are organized showing the base emission rates from MOBILE6.2 on a gram/mile basis. The adjusted emissions rates on a grams/mile basis are shown across the top of the table in summary form for retrofitted vehicles of model years 1990 - 2006. The main body of the tables shows the annual emission reductions estimated as described above in each year of a retrofit vehicles life beginning with retrofit in 2007.

Those annual estimates can then be brought back to a net present value at a defined discount rate (3 percent) to give a discounted lifetime emission reduction. This result is shown in the second row of the lower half of Tables 12-17. The ratio of the cost for the retrofit technology and the discounted lifetime emissions reductions represents the relative cost per ton reduction for the retrofit technology. These results are shown in the last row of the upper half of tables 12 - 17. Because vehicles retrofitted at different ages will have different lifetime emission reductions, we have made estimates for retrofits for various vehicle model years as if the vehicles were retrofitted in calendar year 2007. Hence a 2006 model year vehicle retrofitted in model year 2007 would be one year old, and a 2001 model year vehicle retrofitted in model year 2007 would be six years old. Tables 12 - 17 organize the vehicles of different ages by column designating both the model year of the retrofitted vehicle (e.g., 2001) and the age of the vehicle when retrofitted in 2007 (e.g., 6 years old).

II.G.3. Nonroad Emission Reduction Analysis
We have followed a similar process for the 250 hp bulldozer, using inputs from the NONROAD inventory model and the simplifying assumptions described earlier in this paper. EPA has developed a retrofit modeling module within NMIM that will enable states and other interested parties to directly estimate the emission reduction potential of nonroad retrofits.

III. RESULTS

Table 18 below, summarizes the range of cost effectiveness ratios we estimated for the selected retrofit cases in this paper. As noted previously, these estimates represent a nominal projection of the future cost per ton of emission reduction. These cost effectiveness estimates have not factored in the co-benefits from reducing other pollutants such as HC. The cost effectiveness of retrofitted programs can vary significantly depending on a number of factors, including actual annual average activity (i.e., annual vehicle miles traveled for highway or annual operating hours for nonroad).

Table 18 Summary of Cost Effectiveness for Various Diesel PM Retrofit Scenarios

Vehicle	Retrofit Technology	Range of $/ton PM Emission Reduced	
School Bus	DOC	$12,000	$49,100
	CDPF	$12,400	$50,500
Class 6&7 Truck	DOC	$27,600	$67,900
	CDPF	$28,400	$69,900
Class 8b Truck	DOC	$11,100	$40,600
	CDPF	$12,100	$44,100
250 hp Bulldozer	DOC	$18,100	$49,700
	CDPF	n/a	n/a

The results summarized in Table 18 can be compared to similar estimates for other EPA programs targeted at reducing diesel particulate matter. For example, EPA's Urban Bus Retrofit and Rebuild program of $31,500/ton, EPA's 2007 Heavy-Duty diesel emission standards of $14,200/ton, for and EPA's Nonroad Tier 4 emission standards of $11,200/ton.[10]

The results summarized in Table 18 above and given in more detail in Tables 12 - 17 are characterized by increasing cost per ton of emission reduction for the retrofit of older vehicles in comparison to newer vehicles. This characteristic is to be expected as older vehicles will have a shorter remaining lifetime and hence lower remaining emissions (or emission reductions) prior to vehicle scrappage. In some cases, the cost per ton of emission reductions decreases with older vehicles because of older vehicles' relatively high emissions level. That is, retrofitting an emission control technology on an older engine that, due to historically more lenient emission standards has higher emissions, may lead to a larger emission reduction for the same retrofit cost. This benefit from retrofitting older dirtier vehicles is offset by the shorter remaining life of the older vehicles.

IV. CONCLUSIONS

Our analysis demonstrates that diesel retrofit strategies can be a cost effective way to reduce air pollution. We calculated that the cost-effectiveness of DOC and CDPF retrofits for school buses, Class 6-8b trucks, and 250 hp bulldozers range from roughly $11,000 to $70,000 per ton of PM reduced, depending on number of factors such as vehicle activity, survival rates, emissions rates, effectiveness of DOCs and CDPFs and their costs. These findings indicate that retrofits of diesel engines can be as cost-effective as recent EPA rule-makings to address diesel particulate matter, such as the 2007 Heavy-Duty rule and the Nonroad Tier 4 standards which EPA estimates will cost $14,200/ton of PM reduced and $11,200/ton of PM reduced, respectively.

It is important to note that, while we based our cost effectiveness estimates on robust and recent data sources, there is a significant amount of variability in both the costs and the emissions reductions from retrofit technologies in the field. We believe our analysis adequately represents the cost effectiveness of DOC and CDPF retrofits for the average school bus, Class 6-8b truck, and 250 hp bulldozer, but the cost-effectiveness of retrofits for specific engines and vehicle fleets may differ in certain situations.

EPA has developed a module as part of the National Mobile Inventory Model that will allow users to predict the impact of retrofitting their particular fleets. The new module will be able to generate national, county-level, or fleet-specific mobile source emissions inventories and then use these inventories to estimate emissions reductions from retrofit technologies.

Contact:
Carl Wick
U.S. EPA - Office of Transportation and Air Quality. E-mail: wick.carl@epa.gov

Table 1. Average Annual Mileage Accumulation (Curve Fit Data)[1]

| Vehicle Age | HDDV | | | | | | HDDB | |
	2B 8501-10000	3 10001-14000	4-5 14001-19500	6-7 19501-33000	8A 33001-60000	8B >60000	S.BUS ANY WGT.	T.BUS ANY WGT.
1	27137	32751	30563	40681	87821	124208	(a)	45171
2	24831	28984	28622	36872	78257	112590		43731
3	22721	25650	26805	33420	69735	102060		42337
4	20791	22699	25103	30291	62141	92514		40987
5	19024	20088	23509	27455	55374	83861		39681
6	17407	17778	22016	24885	49343	76017		38416
7	15928	15733	20618	22555	43970	68907		37191
8	14575	13923	19309	20443	39181	62462		36005
9	13336	12321	18083	18529	34915	56620		34857
10	12203	10904	16935	16795	31112	51324		33746
11	11166	9650	15860	15222	27724	46523		32670
12	10217	8540	14853	13797	24705	42172		31629
13	9349	7557	13910	12505	22015	38228		30620
14	8555	6688	13026	11335	19617	34652		29644
15	7828	5919	12199	10273	17481	31411		28699
16	7163	5238	11425	9312	15577	28473		27784
17	6554	4635	10699	8440	13881	25810		26898
18	5997	4102	10020	7650	12369	23396		26041
19	5488	3630	9384	6933	11022	21208		25211
20	5021	3213	8788	6284	9822	19224		24407
21	4595	2843	8230	5696	8752	17426		23629
22	4204	2516	7707	5163	7799	15796		22875
23	3847	2227	7218	4679	6950	14319		22146
24	3520	1971	6760	4241	6193	12979		21440
25	3221	1744	6331	3844	5518	11765		20757
26	2947	1543	5929	3484	4918	10665		20095
27	2697	1366	5552	3158	4382	9667		19454
28	2468	1209	5200	2862	3905	8763		18834
29	2258	1070	4869	2594	3480	7944		18234
30	2066	947	4560	2352	3101	7201		17652

HDDV Heavy duty diesel vehicle
HDDB Heavy duty diesel bus

(a)   Average school bus mileage for all ages = 9.939

---

[1] Fleet Characterization Data for MOBILE6: Development and Use of Age Distributions, Average Annual Mileage Accumulation Rates, and Projected Vehicle Counts, Table 6 page 16, EPA420-R-01-047, September 2001 (www.epa.gov/otaq/models/mobile6/r01047.pdf).

Table 2. Transportation Energy Data Book 1980 Model Year Heavy-Duty Survival Rate

Age	Survival Rate
0	1.00
1	1.00
2	1.00
3	1.00
4	0.99
5	0.97
6	0.95
7	0.92
8	0.89
9	0.86
10	0.83
11	0.79
12	0.75
13	0.72
14	0.68
15	0.64
16	0.60
17	0.56
18	0.52
19	0.48
20	0.44
21	0.41
22	0.37
23	0.34
24	0.31
25	0.28
26	0.25
27	0.22
28	0.20
29	0.18
30	0.16

Table 3. Heavy-Duty Diesel Class 8 Chassis Dynamometer Test Data Set

Vehicle Number	Model Year	Cycle	Number of Tests	Fuel Type
1	1997	OCRTC2	6	Federal Grade 2 Diesel
2	1997	OCRTC2	6	Federal Grade 2 Diesel
3	1997	NYGTC3	5	Federal Grade 2 Diesel
4	1997	NYGTC3	6	Federal Grade 2 Diesel
5	1982	5 Mile	3	Federal Grade 2 Diesel
		CSHVR	4	Federal Grade 2 Diesel
		TEST_D	2	Federal Grade 2 Diesel
		WVU-5P	10	Federal Grade 2 Diesel
6	1992	CSHVR	3	California
		HIWAY	1	California
7	1997	CSHVR	1	California
		HIWAY	1	California
8	1985	5 Mile	4	Federal Grade 2 Diesel
		TEST_D	13	Federal Grade 2 Diesel
9	1994	CSHVR	2	California
		HIWAY	2	California
10	1998	CSHVR	3	California
11	1998	CSHVR	3	California
12	1998	CSHVR	3	California
13	1998	CSHVR	3	California
14	1996	3CBD	3	Federal Grade 2 Diesel
		WHM	3	Federal Grade 2 Diesel
15	1997	3CBD	3	Federal Grade 2 Diesel
		WHM	3	Federal Grade 2 Diesel
16	1997	3CBD	3	Federal Grade 2 Diesel
		WHM	3	Federal Grade 2 Diesel
17	1996	2-5 Mile	11	Federal Grade 2 Diesel
		CSHVR	12	Federal Grade 2 Diesel
18	1995	CSHVR	6	California
19	1996	2-5 Mile	3	California
		CSHVR	9	California
20	1996	2-5 Mile	5	California
		CSHVR	6	California
21	1991	14R	1	Federal Grade 2 Diesel
		CBD	10	Federal Grade 2 Diesel
22	1991	CBD	7	Federal Grade 2 Diesel
23	1991	CBD	19	Federal Grade 2 Diesel
24	1991	14C	1	Federal Grade 2 Diesel
		14R	4	Federal Grade 2 Diesel
		CBD	15	Federal Grade 2 Diesel
		CBD-RT	1	Federal Grade 2 Diesel
25	1991	14R	2	Federal Grade 2 Diesel
		CBD	5	Federal Grade 2 Diesel
26	2000	CSHVR	1	California
		HIWAY	1	California
27	1999	CSHVR	2	California
		HIWAY	1	California
28	1998	CSHVR	2	California

Vehicle Number	Model Year	Cycle	Number of Tests	Fuel Type
		HIWAY	1	California
29	1998	2-5 Mile	8	California
		CSHVR	7	California
		HVDUTY	1	California
30	1998	CSHVR	3	California
31	1998	CSHVR	6	California
32	1998	2-5 Mile	3	California
		CSHVR	4	California
33	1998	CSHVR	3	California
34	1998	CSHVR	6	California
35	1998	CSHVR	4	California
36	1998	2-5 Mile	3	California
		CSHVR	4	California
37	1998	CSHVR	3	California
38	1998	20_mph	1	Federal Grade 2 Diesel
		30_mph	2	Federal Grade 2 Diesel
		40_mph	5	Federal Grade 2 Diesel
		5 Mile	4	Federal Grade 2 Diesel
		CSCYC	1	Federal Grade 2 Diesel
		CSHVR	15	Federal Grade 2 Diesel
		TEST_D	1	Federal Grade 2 Diesel
		WVU-5P	3	Federal Grade 2 Diesel
		YARD	1	Federal Grade 2 Diesel
39	1992	CSHVR	2	California
		HIWAY	1	California
			315	

Table 4. Test Cycle Descriptions of Class 8 Heavy-Duty Diesel Chassis Dynamometer Tests

Cycle	Cycle Description	Number of Vehicles	Model Year Range	Number of Tests
20_MPH	20 mile per hour steady state driving	1	1998	1
30_MPH	30 mile per hour steady state driving	1	1998	2
40_MPH	40 mile per hour steady state driving	1	1998	5
5 Mile	Heavy-Duty vehicle drive cycle over 5 miles.	3	1982-1998	11
2-5MIL	The 5MILE Heavy-Duty drive cycle-twice.	6	1996-1998	33
CBD	Central Business District	5	1991	56
CBD-RT	Routized CBD	1	1991	1
3CBD	Triple Central Business District	3	1996-1997	9
14C	Modified CBD	1	1991	1
14R	Modified and routized CBD	3	1991	7
CSCYC	City Suburban Cycle	1	1998	1
CSHVR	Heavy-Duty vehicle drive cycle.	24	1992-2000	98
CSHVR	Heavy-Duty vehicle drive cycle.	2	1982-1998	19
HIWAY		7	1992-2000	8
HVDUTY		1	1998	1
NYGTC3	Triple Length New York Garbage Truck Cycle	2	1997	11
OCRTC2	Orange County Refuse Truck Cycle-twice	2	1997	12
TEST_D	UDD for Heavy-Duty Vehicles	3	1982-1998	16
WHM		3	1996-1997	9
WVU-5P		2	1982-1998	13
YARD		1	1998	1
				315

Table 5. Emission Rates, and Ratios of Chassis Dynamometer Emissions to Engine Dynamometer Emissions by Model Year

| Model Year | New MSOD Data | | | | MOBILE6.2 | Ratios | |
| | D2 fuel | | CARB fuel | | | D2 fuel | CARB fuel |
	N	PM10(g/mi)	N	PM10(g/mi)	PM10(g/mi)	Ratio	Ratio
1980					2.09		
1981					2.09		
1982	19	3.95			2.01	1.97	
1983					2.00		
1984					2.00		
1985	17	3.12			1.99	1.57	
1986					1.98		
1987					2.06		
1988					1.75		
1989					1.73		
1990					1.18		
1991	65	3.82			0.64	5.97	
1992			7	0.97	0.63		1.54
1993					0.63		
1994			4	0.57	0.23		2.45
1995			6	0.81	0.23		3.53
1996	6	0.58	46	0.75	0.23	2.57	3.31
1997	35	1.06	2	0.245	0.23	4.69	1.08
1998	33	0.51	70	0.37	0.23	2.27	1.63
1999			3	0.71	0.23		3.12
2000			2	0.46	0.23		2.03
2001					0.23		
2002					0.23		
2003					0.23		
2004					0.23		
Totals:	175		140		Weighted Ratio =		2.3

Table 7. Calendar Year 2007 Estimated Retrofit Costs for Combined Class 6&7 and School Buses, and for Class 8b

Cost Component	Diesel Oxidation Catalyst (DOC)	Catalyzed Diesel Particulate Filter (CDPF)
Substrate/Coating/Canning	$260	$1,920
Additional exhaust tubing and mounting hardware	$87	$300
Datalogging and testing for CDPF regeneration	-	$100
Installation	$193	$193
Class 6-7 and School Buses Total (2 significant figures)	$540	$2,500
Ratio Class 6&7 to Class 8b	13/8 times	13/8 times
Class 8b Retrofit Cost (2 significant figures)	$880	$4,100
Class 8b Maintenance Cost		$208
Total Class 8b Cost (2 significant figures)		$4,300

Table 8. Retrofit Survival Fractions as a Function of Vehicle Age at Time of Retrofit

| Vehicle Age From New | Scrappage Table Survival Fraction (based on 1980 vehicles) | | | | | | | | | | | | | | | | | |
| | New | 1 | 2 | 3 | 4 | 5 | 6 | 7 | 8 | 9 | 10 | 11 | 12 | 13 | 14 | 15 | 16 | 17 |
Years	(survival %)	(survival %)	(survival %)	(survival %)	(survival %)	(survival %)	(survival %)	(survival %)	(survival %)	(survival %)	(survival %)	(survival %)	(survival %)	(survival %)	(survival %)	(survival %)	(survival %)	(survival %)
1	100%	0%	0%	0	0	0	0	0	0	0	0	0	0	0	0	0	0	0
2	100%	100%	0%	0	0	0	0	0	0	0	0	0	0	0	0	0	0	0
3	100%	100%	100%	0	0	0	0	0	0	0	0	0	0	0	0	0	0	0
4	98.5%	98.5%	98.5%	100.0%	0	0	0	0	0	0	0	0	0	0	0	0	0	0
5	96.7%	96.7%	96.7%	98.2%	100.0%	0	0	0	0	0	0	0	0	0	0	0	0	0
6	94.5%	94.5%	94.5%	96.0%	97.8%	100.0%	0	0	0	0	0	0	0	0	0	0	0	0
7	92.0%	92.0%	92.0%	93.5%	95.3%	97.5%	100.0%	0	0	0	0	0	0	0	0	0	0	0
8	89.1%	89.1%	89.1%	90.6%	92.4%	94.6%	97.1%	100.0%	0	0	0	0	0	0	0	0	0	0
9	86.0%	86.0%	86.0%	87.5%	89.3%	91.5%	94.0%	96.9%	100.0%	0	0	0	0	0	0	0	0	0
10	82.7%	82.7%	82.7%	84.2%	86.0%	88.2%	90.7%	93.6%	96.7%	100.0%	0	0	0	0	0	0	0	0
11	79.1%	79.1%	79.1%	80.6%	82.4%	84.6%	87.1%	90.0%	93.1%	96.4%	100.0%	0	0	0	0	0	0	0
12	75.4%	75.4%	75.4%	76.9%	78.7%	80.9%	83.4%	86.3%	89.4%	92.7%	96.3%	100.0%	0	0	0	0	0	0
13	71.6%	71.6%	71.6%	73.1%	74.9%	77.1%	79.6%	82.5%	85.6%	88.9%	92.5%	96.2%	100.0%	0	0	0	0	0
14	67.7%	67.7%	67.7%	69.2%	71.0%	73.2%	75.7%	78.6%	81.7%	85.0%	88.6%	92.3%	96.1%	100.0%	0	0	0	0
15	63.7%	63.7%	63.7%	65.2%	67.0%	69.2%	71.7%	74.6%	77.7%	81.0%	84.6%	88.3%	92.1%	96.0%	100.0%	0	0	0
16	59.7%	59.7%	59.7%	61.2%	63.0%	65.2%	67.7%	70.6%	73.7%	77.0%	80.6%	84.3%	88.1%	92.0%	96.0%	100.0%	0	0
17	55.7%	55.7%	55.7%	57.2%	59.0%	61.2%	63.7%	66.6%	69.7%	73.0%	76.6%	80.3%	84.1%	88.0%	92.0%	96.0%	100.0%	0
18	51.8%	51.8%	51.8%	53.3%	55.1%	57.3%	59.8%	62.7%	65.8%	69.1%	72.7%	76.4%	80.2%	84.1%	88.1%	92.1%	96.1%	100.0%
19	47.9%	47.9%	47.9%	49.4%	51.2%	53.4%	55.9%	58.8%	61.9%	65.2%	68.8%	72.5%	76.3%	80.2%	84.2%	88.2%	92.2%	96.1%
20	44.2%	44.2%	44.2%	45.7%	47.5%	49.7%	52.2%	55.1%	58.2%	61.5%	65.1%	68.8%	72.6%	76.5%	80.5%	84.5%	88.5%	92.4%
21	40.6%	40.6%	40.6%	42.1%	43.9%	46.1%	48.6%	51.5%	54.6%	57.9%	61.5%	65.2%	69.0%	72.9%	76.9%	80.9%	84.9%	88.8%
22	37.1%	37.1%	37.1%	38.6%	40.4%	42.6%	45.1%	48.0%	51.1%	54.4%	58.0%	61.7%	65.5%	69.4%	73.4%	77.4%	81.4%	85.3%
23	33.7%	33.7%	33.7%	35.2%	37.0%	39.2%	41.7%	44.6%	47.7%	51.0%	54.6%	58.3%	62.1%	66.0%	70.0%	74.0%	78.0%	81.9%
24	30.6%	30.6%	30.6%	32.1%	33.9%	36.1%	38.6%	41.5%	44.6%	47.9%	51.5%	55.2%	59.0%	62.9%	66.9%	70.9%	74.9%	78.8%
25	27.6%	27.6%	27.6%	29.1%	30.9%	33.1%	35.6%	38.5%	41.6%	44.9%	48.5%	52.2%	56.0%	59.9%	63.9%	67.9%	71.9%	75.8%
26	24.8%	24.8%	24.8%	26.3%	28.1%	30.3%	32.8%	35.7%	38.8%	42.1%	45.7%	49.4%	53.2%	57.1%	61.1%	65.1%	69.1%	73.0%
27	22.2%	22.2%	22.2%	23.7%	25.5%	27.7%	30.2%	33.1%	36.2%	39.5%	43.1%	46.8%	50.6%	54.5%	58.5%	62.5%	66.5%	70.4%
28	19.8%	19.8%	19.8%	21.3%	23.1%	25.3%	27.8%	30.7%	33.8%	37.1%	40.7%	44.4%	48.2%	52.1%	56.1%	60.1%	64.1%	68.0%
29	17.6%	17.6%	17.6%	19.1%	20.9%	23.1%	25.6%	28.5%	31.6%	34.9%	38.5%	42.2%	46.0%	49.9%	53.9%	57.9%	61.9%	65.8%
30	15.5%	15.5%	15.5%	17.0%	18.8%	21.0%	23.5%	26.4%	29.5%	32.8%	36.4%	40.1%	43.9%	47.8%	51.8%	55.8%	59.8%	63.7%

22

Table 9. Annual VMT for Class 6&7 Trucks Weighted by the Survival Fraction from the Age at Retrofit

Vehicle Age	6-7 total mileage	New	1 year old	2 year old	3 year old	4 year old	5 year old	6 year old	7 year old	8 year old	9 year old	10 year old	11 year old	12 year old	13 year old	14 year old	15 year old	16 year old	17 year old
							Weight Class 6&7 (19,501-33,000 lbs) VMT Table												
1	40,681	40,681	0	0	0	0	0	0	0	0	0	0	0	0	0	0	0	0
2	36,872	36,872	36,872	0	0	0	0	0	0	0	0	0	0	0	0	0	0	0
3	33,420	33,420	33,420	33,400	0	0	0	0	0	0	0	0	0	0	0	0	0	0
4	30,291	29,837	29,837	29,837	30,291	0	0	0	0	0	0	0	0	0	0	0	0	0
5	27,455	26,549	26,549	26,549	26,981	27,455	0	0	0	0	0	0	0	0	0	0	0	0
6	24,886	23,516	23,516	23,516	23,880	24,338	24,886	0	0	0	0	0	0	0	0	0	0	0
7	22,555	20,751	20,751	20,751	21,099	21,486	21,991	22,555	0	0	0	0	0	0	0	0	0	0
8	20,443	18,215	18,215	18,215	18,521	18,889	19,339	19,880	20,443	0	0	0	0	0	0	0	0	0
9	18,529	15,935	15,935	15,935	16,213	16,546	16,954	17,417	17,995	18,529	0	0	0	0	0	0	0	0
10	16,755	13,889	13,889	13,889	14,141	14,444	14,813	15,233	15,720	16,241	16,785	0	0	0	0	0	0	0
11	15,222	12,041	12,041	12,041	12,269	12,543	12,878	13,288	13,700	14,172	14,674	15,222	0	0	0	0	0	0
12	13,797	10,403	10,403	10,403	10,610	10,888	11,162	11,507	11,907	12,336	12,790	13,287	13,797	0	0	0	0	0
13	12,505	8,954	8,954	8,954	9,141	9,366	9,641	9,954	10,317	10,704	11,117	11,557	12,030	12,505	0	0	0	0
14	11,335	7,674	7,674	7,674	7,844	8,048	8,237	8,581	8,909	9,261	9,635	10,043	10,462	10,883	11,335	0	0	0
15	10,273	6,544	6,544	6,544	6,688	6,883	7,109	7,356	7,664	7,992	8,321	8,691	9,071	9,461	9,852	10,273	0	0
16	9,312	5,599	5,599	5,599	5,699	5,867	6,071	6,304	6,574	6,863	7,170	7,505	7,860	8,204	8,557	8,940	9,312	0
17	8,440	4,701	4,701	4,701	4,828	4,990	5,165	5,376	5,621	5,883	6,161	6,456	6,777	7,088	7,427	7,765	8,102	8,440
18	7,650	3,963	3,963	3,963	4,077	4,215	4,383	4,575	4,797	5,034	5,286	5,552	5,845	6,135	6,434	6,740	7,046	7,332
19	6,933	3,321	3,321	3,321	3,425	3,560	3,702	3,876	4,077	4,232	4,520	4,770	5,006	5,280	5,560	5,833	6,115	6,382
20	6,284	2,778	2,778	2,778	2,872	2,996	3,123	3,280	3,462	3,657	3,865	4,091	4,323	4,552	4,807	5,059	5,310	5,561
21	5,696	2,313	2,313	2,313	2,398	2,501	2,606	2,768	2,933	3,110	3,238	3,503	3,714	3,930	4,152	4,380	4,608	4,836
22	5,163	1,915	1,915	1,915	1,993	2,086	2,199	2,339	2,478	2,668	2,809	2,995	3,186	3,382	3,583	3,790	3,996	4,203
23	4,679	1,577	1,577	1,577	1,647	1,731	1,834	1,951	2,087	2,232	2,386	2,556	2,728	2,906	3,098	3,275	3,462	3,660
24	4,241	1,298	1,298	1,298	1,361	1,438	1,531	1,637	1,760	1,891	2,031	2,184	2,341	2,502	2,668	2,837	3,007	3,177
25	3,844	1,061	1,061	1,061	1,119	1,188	1,272	1,368	1,480	1,599	1,726	1,864	2,007	2,153	2,303	2,446	2,610	2,764
26	3,484	864	864	864	916	979	1,056	1,143	1,244	1,362	1,467	1,592	1,721	1,853	1,999	2,123	2,268	2,407
27	3,158	701	701	701	748	805	875	954	1,046	1,143	1,247	1,361	1,478	1,588	1,721	1,847	1,974	2,100
28	2,862	557	557	557	610	661	724	795	879	957	1,052	1,166	1,271	1,379	1,491	1,606	1,720	1,835
29	2,594	457	457	457	496	542	589	664	739	820	905	999	1,086	1,193	1,294	1,398	1,502	1,606
30	2,352	365	365	365	400	442	494	553	621	694	771	866	943	1,033	1,124	1,218	1,312	1,406

Continued (17 year old column values):
Vehicle Age	17 year old
18	7,650
19	6,663
20	5,806
21	5,058
22	4,404
23	3,832
24	3,342
25	2,934
26	2,543
27	2,223
28	1,996
29	1,707
30	1,488

23

Table 10. Annual VMT for School Buses weighted by the Survival Fraction from the Age at Retrofit

School Bus VMT Table

Vehicle Age	School Bus VMT	New	1 year old	2 year old	3 year old	4 year old	5 year old	6 year old	7 year old	8 year old	9 year old	10 year old	11 year old	12 year old	13 year old	14 year old	15 year old	16 year old	17 year old
	VMT	176,671	168,723	160,537	155,747	151,435	147,805	144,600	142,007	139,565	137,243	135,208	133,037	130,704	128,180	125,433	122,275	118,681	114,488
1	13,248	13,248	0	0	0	0	0	0	0	0	0	0	0	0	0	0	0	0	0
2	13,248	13,248	13,248	0	0	0	0	0	0	0	0	0	0	0	0	0	0	0	0
3	13,248	13,248	13,248	13,248	0	0	0	0	0	0	0	0	0	0	0	0	0	0	0
4	13,248	13,049	13,049	13,049	13,248	0	0	0	0	0	0	0	0	0	0	0	0	0	0
5	13,248	12,811	12,811	12,811	13,010	13,248	0	0	0	0	0	0	0	0	0	0	0	0	0
6	13,248	12,519	12,519	12,519	12,718	12,957	13,248	0	0	0	0	0	0	0	0	0	0	0	0
7	13,248	12,188	12,188	12,188	12,387	12,625	12,917	13,248	0	0	0	0	0	0	0	0	0	0	0
8	13,248	11,804	11,804	11,804	12,003	12,241	12,533	12,864	13,248	0	0	0	0	0	0	0	0	0	0
9	13,248	11,393	11,393	11,393	11,592	11,830	12,122	12,453	12,837	13,248	0	0	0	0	0	0	0	0	0
10	13,248	10,956	10,956	10,956	11,155	11,393	11,685	12,016	12,400	12,811	13,248	0	0	0	0	0	0	0	0
11	13,248	10,479	10,479	10,479	10,678	10,916	11,208	11,539	11,923	12,334	12,771	13,248	0	0	0	0	0	0	0
12	13,248	9,989	9,989	9,989	10,188	10,426	10,718	11,049	11,433	11,844	12,281	12,758	13,248	0	0	0	0	0	0
13	13,248	9,486	9,486	9,486	9,684	9,923	10,214	10,545	10,930	11,340	11,777	12,254	12,745	13,248	0	0	0	0	0
14	13,248	8,969	8,969	8,969	9,168	9,406	9,696	10,029	10,413	10,824	11,261	11,738	12,228	12,731	13,248	0	0	0	0
15	13,248	8,439	8,439	8,439	8,638	8,876	9,168	9,499	9,883	10,294	10,731	11,208	11,698	12,201	12,718	13,248	0	0	0
16	13,248	7,909	7,909	7,909	8,108	8,346	8,638	8,969	9,353	9,764	10,201	10,678	11,168	11,671	12,188	12,718	13,248	0	0
17	13,248	7,379	7,379	7,379	7,578	7,816	8,108	8,439	8,823	9,234	9,671	10,148	10,638	11,142	11,658	12,188	12,718	13,248	0
18	13,248	6,862	6,862	6,862	7,061	7,300	7,591	7,922	8,306	8,717	9,154	9,631	10,121	10,625	11,142	11,671	12,201	12,731	13,248
19	13,248	6,346	6,346	6,346	6,545	6,783	7,074	7,406	7,790	8,201	8,638	9,115	9,605	10,108	10,625	11,155	11,685	12,215	12,731
20	13,248	5,856	5,856	5,856	6,054	6,293	6,584	6,915	7,300	7,710	8,148	8,624	9,115	9,618	10,135	10,665	11,195	11,724	12,241
21	13,248	5,379	5,379	5,379	5,577	5,816	6,107	6,439	6,823	7,233	7,671	8,148	8,638	9,141	9,658	10,188	10,718	11,248	11,764
22	13,248	4,915	4,915	4,915	5,114	5,352	5,644	5,975	6,359	6,770	7,207	7,684	8,174	8,677	9,194	9,724	10,254	10,784	11,301
23	13,248	4,465	4,465	4,465	4,663	4,902	5,193	5,524	5,909	6,319	6,756	7,233	7,724	8,227	8,744	9,274	9,804	10,333	10,850
24	13,248	4,054	4,054	4,054	4,253	4,491	4,783	5,114	5,498	5,909	6,346	6,823	7,313	7,816	8,333	8,863	9,393	9,923	10,439
25	13,248	3,656	3,656	3,656	3,855	4,094	4,385	4,716	5,100	5,511	5,948	6,425	6,915	7,419	7,936	8,465	8,995	9,525	10,042
26	13,248	3,286	3,286	3,286	3,484	3,723	4,014	4,345	4,730	5,140	5,577	6,054	6,545	7,048	7,565	8,095	8,624	9,154	9,671
27	13,248	2,941	2,941	2,941	3,140	3,378	3,670	4,001	4,385	4,796	5,233	5,710	6,200	6,703	7,220	7,750	8,280	8,810	9,327
28	13,248	2,623	2,623	2,623	2,822	3,060	3,352	3,683	4,067	4,478	4,915	5,392	5,882	6,386	6,902	7,432	7,962	8,492	9,009
29	13,248	2,332	2,332	2,332	2,530	2,769	3,060	3,391	3,776	4,186	4,624	5,100	5,591	6,094	6,611	7,141	7,671	8,201	8,717
30	13,248	2,053	2,053	2,053	2,252	2,491	2,782	3,113	3,497	3,908	4,345	4,822	5,312	5,816	6,333	6,862	7,392	7,922	8,439

24

Table 11. Annual VMT for Class 8b Trucks weighted by the Survival Fraction from the Age at Retrofit

Weight Class8B (>60,000 lbs) VMT Table

Vehicle Age	8b total mileage	New	1 year old	2 year old	3 year old	4 year old	5 year old	6 year old	7 year old	8 year old	9 year old	10 year old	11 year old	12 year old	13 year old	14 year old	15 year old	16 year old	17 year old
		845,176	746,324	656,123	584,624	521,427	466,192	417,456	374,959	337,100	303,366	273,648	246,851	222,664	200,809	181,036	162,919	146,293	130,855
1	124,208	124,208	0	0	0	0	0	0	0	0	0	0	0	0	0	0	0	0	0
2	112,590	112,590	112,590	0	0	0	0	0	0	0	0	0	0	0	0	0	0	0	0
3	102,060	102,060	102,060	102,060	0	0	0	0	0	0	0	0	0	0	0	0	0	0	0
4	92,514	91,126	91,126	91,126	92,514	0	0	0	0	0	0	0	0	0	0	0	0	0	0
5	83,861	81,094	81,094	81,094	82,352	83,861	0	0	0	0	0	0	0	0	0	0	0	0	0
6	76,017	71,836	71,836	71,836	72,976	74,345	76,017	0	0	0	0	0	0	0	0	0	0	0	0
7	68,907	63,394	63,394	63,394	64,428	65,668	67,184	68,907	0	0	0	0	0	0	0	0	0	0	0
8	62,462	55,654	55,654	55,654	56,591	57,715	59,089	60,651	62,462	0	0	0	0	0	0	0	0	0	0
9	56,620	48,693	48,693	48,693	49,543	50,562	51,807	53,223	54,865	56,620	0	0	0	0	0	0	0	0	0
10	51,324	42,445	42,445	42,445	43,215	44,139	45,268	46,551	48,039	49,630	51,324	0	0	0	0	0	0	0	0
11	46,523	36,800	36,800	36,800	37,498	38,335	39,358	40,522	41,871	43,313	44,848	46,523	0	0	0	0	0	0	0
12	42,172	31,798	31,798	31,798	32,430	33,189	34,117	35,171	36,394	37,702	39,093	40,612	42,172	0	0	0	0	0	0
13	38,228	27,371	27,371	27,371	27,945	28,633	29,474	30,429	31,538	32,723	33,985	35,361	36,775	38,228	0	0	0	0	0
14	34,652	23,459	23,459	23,459	23,979	24,603	25,365	26,232	27,236	28,311	29,454	30,702	31,984	33,301	34,652	0	0	0	0
15	31,411	20,009	20,009	20,009	20,480	21,045	21,736	22,522	23,433	24,406	25,443	26,574	27,736	28,930	30,155	31,411	0	0	0
16	28,473	16,998	16,998	16,998	17,425	17,938	18,564	19,276	20,102	20,985	21,924	22,949	24,003	25,085	26,195	27,334	28,473	0	0
17	25,810	14,376	14,376	14,376	14,763	15,228	15,796	16,441	17,189	17,990	18,841	19,770	20,725	21,706	22,713	23,745	24,778	25,810	0
18	23,396	12,119	12,119	12,119	12,470	12,891	13,406	13,991	14,669	15,395	16,167	17,009	17,875	18,764	19,676	20,612	21,548	22,484	23,396
19	21,208	10,159	10,159	10,159	10,477	10,858	11,325	11,855	12,470	13,128	13,828	14,591	15,376	16,182	17,009	17,857	18,705	19,554	20,381
20	19,224	8,497	8,497	8,497	8,785	9,131	9,554	10,035	10,592	11,188	11,823	12,515	13,226	13,957	14,706	15,475	16,244	17,013	17,763
21	17,426	7,075	7,075	7,075	7,336	7,650	8,033	8,469	8,974	9,515	10,090	10,717	11,362	12,024	12,704	13,401	14,098	14,795	15,474
22	15,796	5,860	5,860	5,860	6,097	6,382	6,729	7,124	7,582	8,072	8,593	9,162	9,746	10,346	10,962	11,594	12,226	12,858	13,474
23	14,319	4,826	4,826	4,826	5,040	5,298	5,613	5,971	6,386	6,830	7,303	7,818	8,348	8,892	9,451	10,023	10,596	11,169	11,727
24	12,979	3,972	3,972	3,972	4,166	4,400	4,685	5,010	5,386	5,789	6,217	6,684	7,164	7,658	8,164	8,663	9,202	9,721	10,227
25	11,765	3,247	3,247	3,247	3,424	3,635	3,894	4,188	4,530	4,894	5,282	5,706	6,141	6,588	7,047	7,518	7,988	8,459	8,918
26	10,665	2,645	2,645	2,645	2,805	2,997	3,231	3,498	3,807	4,138	4,490	4,874	5,269	5,674	6,090	6,516	6,943	7,370	7,785
27	9,667	2,146	2,146	2,146	2,291	2,465	2,678	2,919	3,200	3,499	3,818	4,166	4,524	4,892	5,269	5,655	6,042	6,429	6,806
28	8,763	1,735	1,735	1,735	1,867	2,024	2,217	2,436	2,690	2,962	3,251	3,567	3,891	4,224	4,566	4,916	5,267	5,617	5,959
29	7,944	1,398	1,398	1,398	1,517	1,660	1,835	2,034	2,264	2,510	2,772	3,058	3,352	3,654	3,964	4,282	4,600	4,917	5,227
30	7,201	1,116	1,116	1,116	1,224	1,354	1,512	1,692	1,901	2,124	2,362	2,621	2,888	3,161	3,442	3,730	4,018	4,306	4,587

25

Table 12. Class 6&7 Truck DOC Cost per Ton Estimates

Class 6&7 Estimates (2007 Calendar Year Retrofits)

Class 6&7 DOC Cost $540
Class 6&7 DOC Efficiency 20%

	2007	2006	2005	2004	2003	2002	2001	2000	1999	1998	1997	1996	1995	1994	1993	1992	1991	1990
Model Year	2007	2006	2005	2004	2003	2002	2001	2000	1999	1998	1997	1996	1995	1994	1993	1992	1991	1990
Mobile 6 Emission Rate [g/mile]	2.3	0.158	0.158	0.158	0.158	0.158	0.158	0.158	0.158	0.158	0.261	0.264	0.265	0.267	0.516	0.517	0.518	0.775
Adjusted Rate	20%	0.364	0.364	0.364	0.364	0.364	0.364	0.364	0.364	0.364	0.601	0.606	0.611	0.614	1.187	1.190	1.192	1.783
DOC Cost Effectiveness		$27,600	$31,400	$35,200	$39,500	$44,200	$49,300	$54,900	$61,100	$67,900	$45,600	$50,100	$55,100	$60,800	$34,900	$38,700	$43,000	$32,100

Weight Class 6&7 (19,501-33,000 lbs) Annual PM Reductions DOC (tons reduction)

Year	6-7	New	1 year old	2 year old	3 year old	4 year old	5 year old	6 year old	7 year old	8 year old	9 year old	10 year old	11 year old	12 year old	13 year old	14 year old	15 year old	16 year old	17 year old
	Lifetime Tons	—>	0.020	0.017	0.015	0.014	0.012	0.011	0.010	0.009	0.008	0.010	0.011	0.010	0.009	0.015	0.014	0.013	0.017
2007	Annual Tons	—>	0.0030	0.0027	0.0024	0.0022	0.0020	0.0018	0.0016	0.0015	0.0013	0.0020	0.0018	0.0017	0.0015	0.0027	0.0024	0.0022	0.0030
2008	Annual Tons	—>	0.0027	0.0024	0.0022	0.0020	0.0018	0.0016	0.0014	0.0014	0.0012	0.0018	0.0016	0.0016	0.0013	0.0023	0.0021	0.0019	0.0026
2009	Annual Tons	—>	0.0024	0.0021	0.0019	0.0017	0.0015	0.0014	0.0013	0.0011	0.0010	0.0015	0.0014	0.0013	0.0012	0.0020	0.0018	0.0017	0.0023
2010	Annual Tons	—>	0.0021	0.0019	0.0017	0.0015	0.0014	0.0012	0.0011	0.0010	0.0009	0.0013	0.0012	0.0011	0.0010	0.0018	0.0016	0.0015	0.0020
2011	Annual Tons	—>	0.0019	0.0017	0.0015	0.0013	0.0012	0.0011	0.0010	0.0008	0.0007	0.0011	0.0010	0.0009	0.0009	0.0015	0.0014	0.0013	0.0017
2012	Annual Tons	—>	0.0017	0.0015	0.0013	0.0012	0.0010	0.0009	0.0009	0.0007	0.0006	0.0010	0.0009	0.0008	0.0008	0.0013	0.0012	0.0011	0.0015
2013	Annual Tons	—>	0.0015	0.0013	0.0011	0.0010	0.0009	0.0008	0.0007	0.0006	0.0006	0.0009	0.0008	0.0007	0.0007	0.0011	0.0010	0.0010	0.0013
2014	Annual Tons	—>	0.0013	0.0011	0.0010	0.0009	0.0008	0.0007	0.0006	0.0005	0.0005	0.0007	0.0006	0.0006	0.0006	0.0010	0.0009	0.0008	0.0011
2015	Annual Tons	—>	0.0011	0.0010	0.0009	0.0008	0.0007	0.0006	0.0005	0.0005	0.0004	0.0006	0.0005	0.0005	0.0005	0.0009	0.0008	0.0007	0.0010
2016	Annual Tons	—>	0.0010	0.0008	0.0007	0.0006	0.0006	0.0005	0.0004	0.0004	0.0004	0.0005	0.0005	0.0004	0.0004	0.0007	0.0007	0.0006	0.0009
2017	Annual Tons	—>	0.0008	0.0007	0.0006	0.0006	0.0005	0.0004	0.0004	0.0003	0.0003	0.0004	0.0004	0.0004	0.0004	0.0006	0.0006	0.0006	0.0008
2018	Annual Tons	—>	0.0007	0.0006	0.0005	0.0005	0.0004	0.0004	0.0003	0.0003	0.0003	0.0004	0.0004	0.0003	0.0003	0.0006	0.0005	0.0005	0.0007
2019	Annual Tons	—>	0.0006	0.0005	0.0005	0.0004	0.0004	0.0003	0.0003	0.0002	0.0002	0.0003	0.0003	0.0003	0.0003	0.0005	0.0005	0.0004	0.0006
2020	Annual Tons	—>	0.0005	0.0004	0.0004	0.0003	0.0003	0.0003	0.0002	0.0002	0.0002	0.0002	0.0002	0.0002	0.0002	0.0004	0.0004	0.0004	scrapped
2021	Annual Tons	—>	0.0004	0.0005	0.0003	0.0003	0.0002	0.0002	0.0002	0.0002	0.0002	0.0002	0.0002	0.0002	0.0002	0.0004	0.0003	scrapped	
2022	Annual Tons	—>	0.0004	0.0003	0.0003	0.0002	0.0002	0.0002	0.0002	0.0002	0.0001	0.0002	0.0002	0.0002	0.0002	0.0003	scrapped		
2023	Annual Tons	—>	0.0003	0.0003	0.0002	0.0002	0.0002	0.0002	0.0001	0.0001	0.0001	0.0002	0.0002	0.0002	0.0002	scrapped			
2024	Annual Tons	—>	0.0003	0.0002	0.0002	0.0002	0.0001	0.0001	0.0001	0.0001	0.0001	0.0001	0.0001	0.0001	scrapped				
2025	Annual Tons	—>	0.0002	0.0002	0.0002	0.0001	0.0001	0.0001	0.0001	0.0001	0.0001	0.0001	0.0001	scrapped					
2026	Annual Tons	—>	0.0002	0.0002	0.0001	0.0001	0.0001	0.0001	0.0001	0.0001	0.0001	0.0001	scrapped						
2027	Annual Tons	—>	0.0002	0.0001	0.0001	0.0001	0.0001	0.0001	0.0001	0.0001	0.0001	scrapped							
2028	Annual Tons	—>	0.0001	0.0001	0.0001	0.0001	0.0001	0.0001	0.0001	0.0001	scrapped								
2029	Annual Tons	—>	0.0001	0.0001	0.0001	0.0001	0.0001	0.0001	0.0000	scrapped									
2030	Annual Tons	—>	0.0001	0.0001	0.0001	0.0001	0.0001	0.0000	scrapped										
2031	Annual Tons	—>	0.0001	0.0001	0.0000	0.0000	0.0000	scrapped											
2032	Annual Tons	—>	0.0000	0.0000	0.0000	0.0000	scrapped												
2033	Annual Tons	—>	0.0000	0.0000	0.0000	scrapped													
2034	Annual Tons	—>	0.0000	0.0000	scrapped														
2035	Annual Tons	—>	0.0000	scrapped															
2036	Annual Tons	—	scrapped																

Table 13. Class 6&7 Truck CDPF Cost per Ton Estimates

Class 6&7 Estimates (2007 Calendar Year Retrofits)

Class 6&7 CDPF Cost $2,500
Class 6&7 CDPF Efficiency 90%

	2007	2006	2005	2004	2003	2002	2001	2000	1999	1998	1997	1996	1995	1994	1993	1992	1991	1990
Model Year																		
Mobile 6 Emission Rate [g/mile]	2.3	0.158	0.158	0.158	0.158	0.158	0.158	0.158	0.158	0.158	0.261	0.264	0.265	0.267	0.516	0.517	0.518	0.775
Adjusted Rate		0.364	0.364	0.364	0.364	0.364	0.364	0.364	0.364	0.364	0.601	0.606	0.611	0.614	1.187	1.190	1.192	1.783
DOC Cost Effectivness		$28,400	$32,300	$36,200	$40,600	$45,500	$50,800	$56,500	$62,900	$69,900	$46,900	$51,600	$56,700	$62,500	$35,900	$39,800	$44,300	$33,100

Weight Class 6&7 (19,501-33,000 lbs) Annual PM Reductions CDPF (tons reduction)

Year	6-7	New	1 year old	2 year old	3 year old	4 year old	5 year old	6 year old	7 year old	8 year old	9 year old	10 year old	11 year old	12 year old	13 year old	14 year old	15 year old	16 year old	17 year old
	Lifetime Tons		0.088	0.077	0.069	0.062	0.055	0.049	0.044	0.040	0.036	0.053	0.048	0.044	0.040	0.070	0.063	0.056	0.076
2007	Annual Tons	---->	0.0133	0.0121	0.0109	0.0099	0.0090	0.0081	0.0074	0.0067	0.0061	0.0091	0.0083	0.0076	0.0069	0.0121	0.0110	0.0100	0.0135
2008	Annual Tons	---->	0.0121	0.0108	0.0097	0.0088	0.0079	0.0072	0.0065	0.0059	0.0053	0.0079	0.0072	0.0066	0.0060	0.0105	0.0096	0.0087	0.0118
2009	Annual Tons	---->	0.0108	0.0096	0.0086	0.0078	0.0070	0.0063	0.0057	0.0051	0.0046	0.0069	0.0063	0.0057	0.0052	0.0091	0.0083	0.0076	0.0103
2010	Annual Tons	---->	0.0096	0.0085	0.0076	0.0068	0.0061	0.0055	0.0049	0.0044	0.0039	0.0060	0.0054	0.0050	0.0045	0.0079	0.0072	0.0066	0.0089
2011	Annual Tons	---->	0.0085	0.0075	0.0067	0.0060	0.0053	0.0048	0.0043	0.0039	0.0035	0.0052	0.0047	0.0043	0.0039	0.0069	0.0063	0.0057	0.0078
2012	Annual Tons	---->	0.0075	0.0066	0.0058	0.0052	0.0046	0.0041	0.0037	0.0033	0.0030	0.0045	0.0041	0.0037	0.0034	0.0060	0.0054	0.0050	0.0068
2013	Annual Tons	---->	0.0066	0.0057	0.0051	0.0045	0.0040	0.0036	0.0032	0.0029	0.0026	0.0038	0.0035	0.0032	0.0029	0.0052	0.0047	0.0043	0.0059
2014	Annual Tons	---->	0.0057	0.0050	0.0044	0.0039	0.0035	0.0031	0.0028	0.0025	0.0022	0.0033	0.0030	0.0028	0.0025	0.0045	0.0041	0.0038	0.0051
2015	Annual Tons	---->	0.0050	0.0043	0.0038	0.0034	0.0030	0.0027	0.0024	0.0021	0.0019	0.0028	0.0026	0.0024	0.0022	0.0039	0.0035	0.0033	0.0045
2016	Annual Tons	---->	0.0043	0.0038	0.0033	0.0029	0.0026	0.0023	0.0020	0.0018	0.0016	0.0024	0.0022	0.0020	0.0019	0.0033	0.0031	0.0028	0.0039
2017	Annual Tons	---->	0.0038	0.0032	0.0028	0.0025	0.0022	0.0019	0.0017	0.0015	0.0014	0.0021	0.0019	0.0018	0.0016	0.0029	0.0027	0.0025	0.0034
2018	Annual Tons	---->	0.0032	0.0028	0.0024	0.0021	0.0019	0.0016	0.0015	0.0013	0.0012	0.0018	0.0016	0.0015	0.0014	0.0025	0.0023	0.0022	0.0030
2019	Annual Tons	---->	0.0028	0.0024	0.0021	0.0018	0.0016	0.0014	0.0012	0.0011	0.0010	0.0015	0.0014	0.0013	0.0012	0.0022	0.0020	0.0019	0.0026
2020	Annual Tons	---->	0.0024	0.0020	0.0017	0.0015	0.0013	0.0012	0.0011	0.0010	0.0009	0.0010	0.0012	0.0011	0.0010	0.0019	0.0018	0.0017	scrapped
2021	Annual Tons	---->	0.0020	0.0017	0.0015	0.0013	0.0011	0.0010	0.0009	0.0008	0.0007	0.0008	0.0009	0.0008	0.0009	0.0016	0.0015	scrapped	
2022	Annual Tons	---->	0.0017	0.0014	0.0012	0.0011	0.0009	0.0008	0.0008	0.0007	0.0006	0.0007	0.0009	0.0008	0.0008	0.0014	scrapped		
2023	Annual Tons	---->	0.0014	0.0012	0.0010	0.0009	0.0008	0.0007	0.0006	0.0006	0.0005	0.0006	0.0007	0.0007	0.0007	scrapped			
2024	Annual Tons	---->	0.0012	0.0010	0.0009	0.0008	0.0007	0.0006	0.0005	0.0005	0.0004	0.0005	0.0006	0.0006	scrapped				
2025	Annual Tons	---->	0.0010	0.0008	0.0007	0.0006	0.0006	0.0005	0.0004	0.0004	0.0004	0.0004	0.0006	scrapped					
2026	Annual Tons	---->	0.0008	0.0007	0.0005	0.0005	0.0005	0.0004	0.0004	0.0003	0.0003	0.0003	scrapped						
2027	Annual Tons	---->	0.0007	0.0006	0.0005	0.0004	0.0004	0.0003	0.0003	0.0003	0.0003	scrapped							
2028	Annual Tons	---->	0.0006	0.0005	0.0004	0.0004	0.0003	0.0003	0.0003	0.0003	scrapped								
2029	Annual Tons	---->	0.0005	0.0004	0.0003	0.0003	0.0003	0.0002	0.0002	scrapped									
2030	Annual Tons	---->	0.0004	0.0003	0.0003	0.0002	0.0002	0.0002	scrapped										
2031	Annual Tons	---->	0.0003	0.0003	0.0002	0.0002	0.0002	scrapped											
2032	Annual Tons	---->	0.0003	0.0002	0.0002	0.0002	scrapped												
2033	Annual Tons	---->	0.0002	0.0002	0.0002	scrapped													
2034	Annual Tons	---->	0.0002	0.0001	scrapped														
2035	Annual Tons	---->	0.0001	scrapped															
2036	-		scrapped																

Table 14. School Bus DOC Cost per Ton Estimates

School Bus Estimates (2007 Calendar Year Retrofits)

School Bus DOC Cost $540
School Bus DOC Efficiency 20%

Model Year	2007	2006	2005	2004	2003	2002	2001	2000	1999	1998	1997	1996	1995	1994	1993	1992	1991	1990
Mobile 6 Emission Rate [g/mile]		0.158	0.158	0.158	0.158	0.158	0.158	0.158	0.158	0.158	0.261	0.264	0.265	0.267	0.516	0.517	0.518	0.775
Adjusted Rate	2.3	0.364	0.364	0.364	0.364	0.364	0.364	0.364	0.364	0.364	0.601	0.606	0.611	0.614	1.187	1.190	1.192	1.783
DOC Cost Effectiveness		$39,900	$42,000	$43,300	$44,500	$45,600	$46,600	$47,500	$48,300	$49,100	$30,200	$30,400	$30,700	$31,100	$16,500	$16,900	$17,300	$12,000

School Bus Annual PM Reductions DOC (tons reduction)

Calendar Yr (age)	School Bus	New	1 year old	2 year old	3 year old	4 year old	5 year old	6 year old	7 year old	8 year old	9 year old	10 year old	11 year old	12 year old	13 year old	14 year old	15 year old	16 year old	17 year old
	Lifetime Tons		0.014	0.013	0.012	0.012	0.012	0.012	0.011	0.011	0.011	0.018	0.018	0.018	0.017	0.033	0.032	0.031	0.045
2007	Annual Tons	----->	0.0011	0.0011	0.0011	0.0011	0.0011	0.0011	0.0011	0.0011	0.0011	0.0018	0.0018	0.0018	0.0018	0.0035	0.0035	0.0035	0.0052
2008	Annual Tons	----->	0.0011	0.0010	0.0010	0.0010	0.0010	0.0010	0.0010	0.0010	0.0010	0.0017	0.0017	0.0017	0.0017	0.0033	0.0033	0.0033	0.0050
2009	Annual Tons	----->	0.0010	0.0010	0.0010	0.0010	0.0010	0.0010	0.0010	0.0010	0.0010	0.0016	0.0016	0.0016	0.0016	0.0032	0.0032	0.0032	0.0048
2010	Annual Tons	----->	0.0010	0.0010	0.0010	0.0010	0.0010	0.0010	0.0009	0.0009	0.0010	0.0016	0.0016	0.0016	0.0016	0.0031	0.0031	0.0031	0.0046
2011	Annual Tons	----->	0.0010	0.0010	0.0009	0.0009	0.0009	0.0009	0.0009	0.0009	0.0009	0.0015	0.0015	0.0015	0.0015	0.0029	0.0029	0.0030	0.0044
2012	Annual Tons	----->	0.0010	0.0009	0.0009	0.0009	0.0009	0.0009	0.0009	0.0009	0.0009	0.0014	0.0014	0.0014	0.0014	0.0028	0.0028	0.0028	0.0043
2013	Annual Tons	----->	0.0009	0.0009	0.0008	0.0009	0.0008	0.0008	0.0008	0.0008	0.0008	0.0013	0.0014	0.0014	0.0013	0.0027	0.0027	0.0027	0.0041
2014	Annual Tons	----->	0.0009	0.0009	0.0008	0.0008	0.0008	0.0008	0.0008	0.0008	0.0008	0.0013	0.0013	0.0014	0.0013	0.0025	0.0026	0.0026	0.0039
2015	Annual Tons	----->	0.0009	0.0008	0.0008	0.0007	0.0007	0.0008	0.0008	0.0007	0.0007	0.0012	0.0012	0.0012	0.0012	0.0024	0.0025	0.0025	0.0038
2016	Annual Tons	----->	0.0008	0.0008	0.0007	0.0007	0.0007	0.0007	0.0007	0.0007	0.0007	0.0011	0.0012	0.0012	0.0011	0.0023	0.0024	0.0024	0.0037
2017	Annual Tons	----->	0.0008	0.0007	0.0007	0.0007	0.0006	0.0007	0.0007	0.0007	0.0007	0.0011	0.0011	0.0011	0.0011	0.0022	0.0023	0.0023	0.0035
2018	Annual Tons	----->	0.0008	0.0007	0.0006	0.0006	0.0006	0.0006	0.0006	0.0006	0.0006	0.0010	0.0010	0.0011	0.0010	0.0021	0.0022	0.0022	0.0034
2019	Annual Tons	----->	0.0007	0.0007	0.0006	0.0006	0.0006	0.0006	0.0006	0.0006	0.0006	0.0010	0.0010	0.0010	0.0010	0.0020	0.0021	0.0022	0.0033
2020	Annual Tons	----->	0.0007	0.0006	0.0006	0.0006	0.0005	0.0005	0.0005	0.0005	0.0005	0.0009	0.0009	0.0009	0.0010	0.0019	0.0020	0.0021	scrapped
2021	Annual Tons	----->	0.0006	0.0006	0.0005	0.0005	0.0005	0.0005	0.0005	0.0005	0.0005	0.0008	0.0009	0.0009	0.0009	0.0019	0.0019	scrapped	
2022	Annual Tons	----->	0.0006	0.0005	0.0005	0.0005	0.0005	0.0005	0.0005	0.0005	0.0005	0.0008	0.0008	0.0009	0.0009	0.0018	scrapped		
2023	Annual Tons	----->	0.0005	0.0005	0.0005	0.0005	0.0005	0.0004	0.0004	0.0004	0.0005	0.0008	0.0008	0.0008	0.0009	scrapped			
2024	Annual Tons	----->	0.0005	0.0005	0.0004	0.0004	0.0004	0.0004	0.0004	0.0004	0.0004	0.0007	0.0007	0.0008	scrapped				
2025	Annual Tons	----->	0.0005	0.0004	0.0004	0.0004	0.0004	0.0004	0.0004	0.0004	0.0004	0.0007	0.0008	scrapped					
2026	Annual Tons	----->	0.0004	0.0004	0.0003	0.0004	0.0004	0.0003	0.0004	0.0004	0.0004	0.0007	scrapped						
2027	Annual Tons	----->	0.0004	0.0003	0.0003	0.0003	0.0003	0.0003	0.0003	0.0004	0.0004	scrapped							
2028	Annual Tons	----->	0.0004	0.0003	0.0003	0.0003	0.0003	0.0003	0.0003	0.0003	scrapped								
2029	Annual Tons	----->	0.0003	0.0003	0.0003	0.0003	0.0003	0.0003	0.0003	scrapped									
2030	Annual Tons	----->	0.0003	0.0003	0.0002	0.0003	0.0003	0.0002	scrapped										
2031	Annual Tons	----->	0.0003	0.0002	0.0002	0.0002	0.0002	scrapped											
2032	Annual Tons	----->	0.0002	0.0002	0.0002	0.0002	scrapped												
2033	Annual Tons	----->	0.0002	0.0002	0.0002	scrapped													
2034	Annual Tons	----->	0.0002	0.0002	scrapped														
2035	Annual Tons	----->	0.0002	scrapped															
2036	Annual Tons	----->	scrapped																

Table 15. School Bus CDPF Cost per Ton Estimates

School Bus Estimates (2007 Calendar Year Retrofits)

School Bus CDPF Cost $2,500
School Bus CDPF Efficiency   90%

	2007	2006	2005	2004	2003	2002	2001	2000	1999	1998	1997	1996	1995	1994	1993	1992	1991	1990
Model Year	2007	2006	2005	2004	2003	2002	2001	2000	1999	1998	1997	1996	1995	1994	1993	1992	1991	1990
Mobile 6 Emission Rate [g/mile]	2.3	0.158	0.158	0.158	0.158	0.158	0.158	0.158	0.158	0.158	0.261	0.264	0.265	0.267	0.516	0.517	0.518	0.775
Adjusted Rate		0.364	0.364	0.364	0.364	0.364	0.364	0.364	0.364	0.364	0.601	0.606	0.611	0.614	1.187	1.190	1.192	1.783
DOC Cost Effectiveness		$41,100	$43,200	$44,500	$45,800	$46,900	$47,900	$48,800	$49,700	$50,500	$31,100	$31,300	$31,600	$32,000	$16,900	$17,300	$17,800	$12,400

School Bus Annual PM Reductions CDPFs (tons reduction)

Vehicle Age	School Bus	New	1 year old	2 year old	3 year old	4 year old	5 year old	6 year old	7 year old	8 year old	9 year old	10 year old	11 year old	12 year old	13 year old	14 year old	15 year old	16 year old	17 year old
	Lifetime Tons		0.061	0.058	0.056	0.055	0.053	0.052	0.051	0.050	0.049	0.080	0.080	0.079	0.078	0.148	0.144	0.140	0.202
2007	Annual Tons	------>	0.0048	0.0048	0.0047	0.0046	0.0045	0.0044	0.0043	0.0041	0.0040	0.0079	0.0080	0.0080	0.0081	0.0156	0.0156	0.0157	0.0234
2008	Annual Tons	------>	0.0048	0.0047	0.0046	0.0045	0.0044	0.0043	0.0041	0.0040	0.0038	0.0076	0.0077	0.0077	0.0077	0.0150	0.0150	0.0150	0.0225
2009	Annual Tons	------>	0.0047	0.0046	0.0045	0.0044	0.0043	0.0041	0.0040	0.0038	0.0036	0.0073	0.0074	0.0074	0.0074	0.0143	0.0144	0.0144	0.0216
2010	Annual Tons	------>	0.0046	0.0045	0.0044	0.0043	0.0041	0.0040	0.0038	0.0036	0.0034	0.0070	0.0070	0.0071	0.0071	0.0137	0.0138	0.0139	0.0208
2011	Annual Tons	------>	0.0045	0.0044	0.0043	0.0041	0.0040	0.0038	0.0036	0.0034	0.0032	0.0067	0.0067	0.0067	0.0068	0.0131	0.0132	0.0133	0.0200
2012	Annual Tons	------>	0.0044	0.0043	0.0041	0.0040	0.0038	0.0036	0.0034	0.0032	0.0030	0.0064	0.0064	0.0064	0.0065	0.0125	0.0126	0.0127	0.0192
2013	Annual Tons	------>	0.0043	0.0041	0.0040	0.0038	0.0036	0.0034	0.0032	0.0030	0.0029	0.0060	0.0061	0.0061	0.0062	0.0120	0.0121	0.0122	0.0185
2014	Annual Tons	------>	0.0041	0.0040	0.0038	0.0036	0.0034	0.0032	0.0030	0.0029	0.0027	0.0057	0.0058	0.0058	0.0059	0.0114	0.0116	0.0117	0.0177
2015	Annual Tons	------>	0.0040	0.0038	0.0036	0.0034	0.0032	0.0030	0.0029	0.0027	0.0025	0.0054	0.0055	0.0055	0.0056	0.0109	0.0111	0.0113	0.0171
2016	Annual Tons	------>	0.0038	0.0036	0.0034	0.0032	0.0030	0.0029	0.0027	0.0025	0.0023	0.0051	0.0053	0.0053	0.0053	0.0104	0.0106	0.0108	0.0165
2017	Annual Tons	------>	0.0036	0.0034	0.0032	0.0030	0.0029	0.0027	0.0025	0.0023	0.0021	0.0049	0.0050	0.0050	0.0051	0.0100	0.0102	0.0104	0.0159
2018	Annual Tons	------>	0.0034	0.0032	0.0030	0.0029	0.0027	0.0025	0.0023	0.0021	0.0019	0.0046	0.0047	0.0047	0.0048	0.0095	0.0098	0.0100	0.0154
2019	Annual Tons	------>	0.0032	0.0030	0.0029	0.0027	0.0025	0.0023	0.0021	0.0019	0.0018	0.0043	0.0045	0.0045	0.0046	0.0091	0.0094	0.0097	0.0149
2020	Annual Tons	------>	0.0030	0.0029	0.0027	0.0025	0.0023	0.0021	0.0019	0.0018	0.0016	0.0041	0.0043	0.0043	0.0044	0.0087	0.0090	0.0094	scrapped
2021	Annual Tons	------>	0.0029	0.0027	0.0025	0.0023	0.0021	0.0019	0.0018	0.0016	0.0015	0.0038	0.0041	0.0041	0.0042	0.0084	0.0087	scrapped	
2022	Annual Tons	------>	0.0027	0.0025	0.0023	0.0021	0.0019	0.0018	0.0016	0.0015	0.0013	0.0036	0.0039	0.0039	0.0040	0.0081	scrapped		
2023	Annual Tons	------>	0.0025	0.0023	0.0021	0.0019	0.0018	0.0016	0.0015	0.0013	0.0012	0.0034	0.0037	0.0037	0.0039	scrapped			
2024	Annual Tons	------>	0.0023	0.0021	0.0019	0.0018	0.0016	0.0015	0.0013	0.0012	0.0011	0.0032	0.0035	0.0035	scrapped				
2025	Annual Tons	------>	0.0021	0.0019	0.0018	0.0016	0.0015	0.0013	0.0012	0.0011	0.0009	0.0030	0.0034	scrapped					
2026	Annual Tons	------>	0.0019	0.0018	0.0016	0.0015	0.0013	0.0012	0.0011	0.0009	0.0008	0.0029	scrapped						
2027	Annual Tons	------>	0.0018	0.0016	0.0015	0.0013	0.0012	0.0011	0.0009	0.0008	0.0007	scrapped							
2028	Annual Tons	------>	0.0016	0.0015	0.0013	0.0012	0.0011	0.0009	0.0008	0.0007	scrapped								
2029	Annual Tons	------>	0.0015	0.0013	0.0012	0.0011	0.0009	0.0008	0.0007	scrapped									
2030	Annual Tons	------>	0.0013	0.0012	0.0011	0.0009	0.0008	0.0007	scrapped										
2031	Annual Tons	------>	0.0012	0.0011	0.0009	0.0008	0.0007	scrapped											
2032	Annual Tons	------>	0.0011	0.0009	0.0008	0.0007	scrapped												
2033	Annual Tons	------>	0.0009	0.0008	0.0007	scrapped													
2034	Annual Tons	------>	0.0008	0.0007	scrapped														
2035	Annual Tons	------>	0.0007	scrapped															
2036	-		scrapped																

29

Table 16. Class 8b Truck DOC Cost per Ton Estimates

Class 8b Estimates (2007 Calendar Year Retrofits)

Class 8b DOC Cost $880
Class 8b DOC Efficiency 20%

Model Year	2007	2006	2005	2004	2003	2002	2001	2000	1999	1998	1997	1996	1995	1994	1993	1992	1991	1990
Mobile 6 Emission Rate [g/mile]		0.209	0.209	0.209	0.209	0.209	0.209	0.209	0.209	0.209	0.209	0.209	0.211	0.213	0.579	0.584	0.589	1.084
Adjusted Rate	2.3	0.482	0.482	0.482	0.482	0.482	0.482	0.482	0.482	0.482	0.482	0.482	0.486	0.489	1.332	1.343	1.356	2.494
DOC Cost Effectiveness		$11,100	$12,600	$14,200	$15,900	$17,800	$19,900	$22,100	$24,600	$27,300	$30,300	$33,600	$37,000	$40,600	$16,600	$18,300	$20,100	$12,200

Weight Class 8b (>60,000 lbs) Annual PM Reductions DOC (tons reduction)

Year	8b	New	1 year old	2 year old	3 year old	4 year old	5 year old	6 year old	7 year old	8 year old	9 year old	10 year old	11 year old	12 year old	13 year old	14 year old	15 year old	16 year old	17 year old
	Lifetime Tons	------>	0.079	0.070	0.062	0.055	0.049	0.044	0.040	0.036	0.032	0.029	0.026	0.024	0.022	0.053	0.048	0.044	0.072
2007	Annual Tons	------>	0.0119	0.0108	0.0098	0.0089	0.0081	0.0073	0.0066	0.0060	0.0054	0.0049	0.0045	0.0041	0.0037	0.0092	0.0084	0.0077	0.0129
2008	Annual Tons	------>	0.0108	0.0097	0.0087	0.0079	0.0071	0.0064	0.0058	0.0053	0.0048	0.0043	0.0039	0.0036	0.0033	0.0080	0.0073	0.0067	0.0112
2009	Annual Tons	------>	0.0097	0.0086	0.0077	0.0070	0.0063	0.0056	0.0051	0.0046	0.0041	0.0038	0.0034	0.0031	0.0028	0.0070	0.0064	0.0058	0.0098
2010	Annual Tons	------>	0.0086	0.0076	0.0068	0.0061	0.0055	0.0049	0.0044	0.0040	0.0036	0.0033	0.0029	0.0027	0.0024	0.0060	0.0055	0.0051	0.0085
2011	Annual Tons	------>	0.0076	0.0067	0.0060	0.0054	0.0048	0.0043	0.0039	0.0035	0.0031	0.0028	0.0025	0.0023	0.0021	0.0052	0.0048	0.0044	0.0074
2012	Annual Tons	------>	0.0067	0.0059	0.0053	0.0047	0.0042	0.0037	0.0033	0.0030	0.0027	0.0024	0.0022	0.0020	0.0018	0.0045	0.0042	0.0038	0.0064
2013	Annual Tons	------>	0.0059	0.0052	0.0046	0.0041	0.0036	0.0032	0.0029	0.0026	0.0023	0.0021	0.0019	0.0017	0.0016	0.0039	0.0036	0.0033	0.0056
2014	Annual Tons	------>	0.0052	0.0045	0.0040	0.0035	0.0031	0.0028	0.0025	0.0022	0.0020	0.0018	0.0016	0.0015	0.0014	0.0034	0.0031	0.0029	0.0049
2015	Annual Tons	------>	0.0045	0.0039	0.0034	0.0030	0.0027	0.0024	0.0021	0.0019	0.0017	0.0015	0.0014	0.0013	0.0012	0.0029	0.0027	0.0025	0.0043
2016	Annual Tons	------>	0.0039	0.0034	0.0030	0.0026	0.0023	0.0020	0.0018	0.0016	0.0015	0.0013	0.0012	0.0011	0.0010	0.0025	0.0024	0.0022	0.0037
2017	Annual Tons	------>	0.0034	0.0029	0.0025	0.0022	0.0020	0.0017	0.0016	0.0014	0.0013	0.0011	0.0010	0.0010	0.0009	0.0022	0.0021	0.0019	0.0033
2018	Annual Tons	------>	0.0029	0.0025	0.0022	0.0019	0.0017	0.0015	0.0013	0.0012	0.0011	0.0010	0.0009	0.0008	0.0008	0.0019	0.0018	0.0017	0.0029
2019	Annual Tons	------>	0.0025	0.0021	0.0018	0.0016	0.0014	0.0013	0.0011	0.0010	0.0009	0.0009	0.0008	0.0007	0.0007	0.0017	0.0016	0.0015	0.0025
2020	Annual Tons	------>	0.0021	0.0018	0.0016	0.0014	0.0012	0.0011	0.0010	0.0009	0.0008	0.0008	0.0007	0.0006	0.0006	0.0014	0.0014	0.0013	scrapped
2021	Annual Tons	------>	0.0018	0.0015	0.0013	0.0012	0.0010	0.0009	0.0008	0.0008	0.0007	0.0007	0.0006	0.0005	0.0005	0.0013	0.0012	scrapped	
2022	Annual Tons	------>	0.0015	0.0013	0.0011	0.0010	0.0009	0.0008	0.0007	0.0007	0.0006	0.0006	0.0005	0.0005	0.0004	0.0011	scrapped		
2023	Annual Tons	------>	0.0013	0.0011	0.0009	0.0008	0.0007	0.0007	0.0006	0.0006	0.0005	0.0005	0.0005	0.0004	0.0004	scrapped			
2024	Annual Tons	------>	0.0011	0.0009	0.0008	0.0007	0.0006	0.0006	0.0005	0.0005	0.0004	0.0004	0.0004	0.0003	scrapped				
2025	Annual Tons	------>	0.0009	0.0008	0.0006	0.0006	0.0005	0.0005	0.0004	0.0004	0.0004	0.0004	0.0003	scrapped					
2026	Annual Tons	------>	0.0008	0.0006	0.0005	0.0005	0.0004	0.0004	0.0003	0.0003	0.0003	0.0003	scrapped						
2027	Annual Tons	------>	0.0006	0.0005	0.0004	0.0004	0.0003	0.0003	0.0003	0.0003	0.0003	scrapped							
2028	Annual Tons	------>	0.0005	0.0004	0.0004	0.0003	0.0003	0.0003	0.0002	0.0002	scrapped								
2029	Annual Tons	------>	0.0004	0.0003	0.0003	0.0003	0.0002	0.0002	0.0002	scrapped									
2030	Annual Tons	------>	0.0003	0.0003	0.0002	0.0002	0.0002	0.0002	scrapped										
2031	Annual Tons	------>	0.0003	0.0002	0.0002	0.0002	0.0002	scrapped											
2032	Annual Tons	------>	0.0002	0.0002	0.0002	0.0001	scrapped												
2033	Annual Tons	------>	0.0002	0.0001	0.0001	scrapped													
2034	Annual Tons	------>	0.0001	0.0001	scrapped														
2035	Annual Tons	------>	0.0001	scrapped															
2036	Annual Tons	-	scrapped																

Table 17. Class 8b Truck CDPF Cost per Ton Estimates

Class 8b Estimates (2007 Calendar Year Retrofits)

Class 8b CDPF Cost $4,300
Class 8b CDPF Efficiency 90%

Model Year	2007	2006	2005	2004	2003	2002	2001	2000	1999	1998	1997	1996	1995	1994	1993	1992	1991	1990
Mobile 6 Emission Rate [g/mile]	0.209	0.209	0.209	0.209	0.209	0.209	0.209	0.209	0.209	0.209	0.209	0.209	0.211	0.213	0.579	0.584	0.589	1.084
Adjusted Rate [g/mile]	0.482	0.482	0.482	0.482	0.482	0.482	0.482	0.482	0.482	0.482	0.482	0.482	0.486	0.489	1.332	1.343	1.356	2.494
DOC Cost Effectiveness [$/ton]		$12,100	$13,700	$15,400	$17,300	$19,300	$21,600	$24,000	$26,700	$29,700	$32,900	$36,500	$40,100	$44,100	$18,000	$19,800	$21,900	$13,300

Weight Class 8b (>60,000 lbs) Annual PM Reductions CDPF (tons reduction)

Year	6-7	New	1 year old	2 year old	3 year old	4 year old	5 year old	6 year old	7 year old	8 year old	9 year old	10 year old	11 year old	12 year old	13 year old	14 year old	15 year old	16 year old	17 year old
	Lifetime Tons	——>	0.356	0.313	0.279	0.249	0.223	0.199	0.179	0.161	0.145	0.131	0.118	0.107	0.097	0.239	0.217	0.197	0.323
2007	Annual Tons	——>	0.0538	0.0487	0.0442	0.0400	0.0363	0.0329	0.0298	0.0270	0.0245	0.0222	0.0201	0.0184	0.0168	0.0415	0.0379	0.0347	0.0578
2008	Annual Tons	——>	0.0487	0.0435	0.0393	0.0355	0.0321	0.0290	0.0262	0.0237	0.0214	0.0194	0.0176	0.0160	0.0146	0.0361	0.0330	0.0302	0.0504
2009	Annual Tons	——>	0.0435	0.0387	0.0348	0.0314	0.0282	0.0254	0.0229	0.0207	0.0187	0.0169	0.0153	0.0139	0.0127	0.0313	0.0287	0.0263	0.0439
2010	Annual Tons	——>	0.0387	0.0343	0.0308	0.0276	0.0247	0.0222	0.0200	0.0180	0.0162	0.0147	0.0132	0.0121	0.0110	0.0272	0.0249	0.0229	0.0383
2011	Annual Tons	——>	0.0343	0.0303	0.0270	0.0241	0.0216	0.0193	0.0174	0.0156	0.0141	0.0127	0.0115	0.0104	0.0095	0.0236	0.0216	0.0199	0.0333
2012	Annual Tons	——>	0.0303	0.0266	0.0237	0.0211	0.0188	0.0168	0.0151	0.0135	0.0121	0.0110	0.0099	0.0090	0.0083	0.0204	0.0188	0.0173	0.0290
2013	Annual Tons	——>	0.0266	0.0232	0.0206	0.0183	0.0163	0.0145	0.0130	0.0117	0.0105	0.0094	0.0085	0.0078	0.0071	0.0177	0.0163	0.0150	0.0253
2014	Annual Tons	——>	0.0232	0.0203	0.0179	0.0158	0.0141	0.0125	0.0112	0.0100	0.0090	0.0081	0.0073	0.0067	0.0062	0.0153	0.0141	0.0131	0.0220
2015	Annual Tons	——>	0.0203	0.0176	0.0155	0.0137	0.0121	0.0108	0.0096	0.0086	0.0077	0.0070	0.0063	0.0058	0.0053	0.0132	0.0123	0.0114	0.0192
2016	Annual Tons	——>	0.0176	0.0152	0.0133	0.0117	0.0104	0.0092	0.0082	0.0073	0.0066	0.0060	0.0054	0.0050	0.0046	0.0115	0.0106	0.0099	0.0168
2017	Annual Tons	——>	0.0152	0.0131	0.0114	0.0100	0.0089	0.0078	0.0070	0.0063	0.0056	0.0051	0.0047	0.0043	0.0040	0.0099	0.0092	0.0086	0.0147
2018	Annual Tons	——>	0.0131	0.0112	0.0098	0.0086	0.0075	0.0067	0.0060	0.0053	0.0048	0.0044	0.0040	0.0037	0.0034	0.0086	0.0080	0.0075	0.0129
2019	Annual Tons	——>	0.0112	0.0096	0.0083	0.0073	0.0064	0.0057	0.0051	0.0045	0.0041	0.0037	0.0034	0.0032	0.0030	0.0075	0.0070	0.0066	0.0113
2020	Annual Tons	——>	0.0096	0.0081	0.0070	0.0062	0.0054	0.0048	0.0043	0.0039	0.0035	0.0032	0.0029	0.0027	0.0026	0.0065	0.0061	0.0058	scrapped
2021	Annual Tons	——>	0.0081	0.0069	0.0060	0.0052	0.0046	0.0040	0.0036	0.0033	0.0030	0.0027	0.0025	0.0024	0.0022	0.0057	0.0054	scrapped	
2022	Annual Tons	——>	0.0069	0.0058	0.0050	0.0044	0.0038	0.0034	0.0030	0.0028	0.0025	0.0023	0.0022	0.0020	0.0019	0.0049	scrapped		
2023	Annual Tons	——>	0.0058	0.0048	0.0042	0.0037	0.0032	0.0029	0.0026	0.0023	0.0021	0.0020	0.0019	0.0018	0.0017	scrapped			
2024	Annual Tons	——>	0.0048	0.0041	0.0035	0.0030	0.0027	0.0024	0.0022	0.0020	0.0018	0.0017	0.0016	0.0015	scrapped				
2025	Annual Tons	——>	0.0041	0.0034	0.0029	0.0025	0.0022	0.0020	0.0018	0.0017	0.0016	0.0015	0.0014	scrapped					
2026	Annual Tons	——>	0.0034	0.0028	0.0024	0.0021	0.0019	0.0017	0.0015	0.0014	0.0013	0.0013	scrapped						
2027	Annual Tons	——>	0.0028	0.0023	0.0020	0.0017	0.0015	0.0014	0.0013	0.0012	0.0011	scrapped							
2028	Annual Tons	——>	0.0023	0.0019	0.0016	0.0014	0.0013	0.0012	0.0011	0.0010	scrapped								
2029	Annual Tons	——>	0.0019	0.0016	0.0013	0.0012	0.0011	0.0010	0.0009	scrapped									
2030	Annual Tons	——>	0.0016	0.0013	0.0011	0.0010	0.0009	0.0008	scrapped										
2031	Annual Tons	——>	0.0013	0.0010	0.0009	0.0008	0.0007	scrapped											
2032	Annual Tons	——>	0.0010	0.0008	0.0007	0.0006	scrapped												
2033	Annual Tons	——>	0.0008	0.0007	0.0006	scrapped													
2034	Annual Tons	——>	0.0007	0.0005	scrapped														
2035	Annual Tons	——>	0.0005	scrapped															
2036	Annual Tons	-	scrapped																

REFERENCES

1. Bobit Publications, "School Bus Fleet 1997 Fact Book", EPA420-R-01-047, September 2001 and Federal Transit Authority 1997 facts.

2. Fleet Characterization Data for MOBILE 6: Development and Use of Age Distributions, Average Annual Mileage Accumulation Rates and Projected Vehicle Counts for Use in MOBILE6, EPA420-P-99-011 April 1999 M6.FLT.007 available on EPA's website at www.epa.gov/otaq/models/mobile6/m6tech.htm

3. Median Life, Annual Activity, and Load Factor Values for Nonroad engine Emissions Modeling, NR-005c (EPA420-P-004-005, April 2004), available at www.epa.gov/otaq/nonrdmdl.htm#techrept

4. Alan Greenspan and Darrel Cohen, "Motor Vehicle Stocks, Scrappage, and Sales", Federal Reserve Board, October 30, 1996 available at http://www.federalreserve.gov/pubs/feds/1996/199640/199640pap.pdf

5. Exhaust and Crankcase Emission Factors for Nonroad Engine Modeling --Compression-Ignition, NR-009c (EPA420-P-04-009, April 2004), available at www.epa.gov/otaq/nonrdmdl.htm#techrept

6. MECA Independent Cost Survey for Emission Control Retrofit Technologies, Manufacturers of Emission Control Association, December 5, 2000 available on EPA's Retrofit Website, www.epa.gov/otaq/retrofit

7. Highway Diesel Progress Review Report 2, March 2004 EPA420-R-04-004 available at www.epa.gov/otaq/highway-diesel/index.htm

8. Nonroad Tier 4 Regulatory Impact Analysis (RIA), (EPA420-R-04-007, May 2004) http://www.epa.gov/nonroad-diesel/2004fr.htm

9. LeTavec, Chuck, et al, "Year-Long Evaluation of Trucks and Buses Equipped with Passive Diesel Particulate Filters," SAE 2002-01-0433.

10. Control of Emissions of Air Pollution From Nonroad Diesel Engines and Fuel, Table IV.D-4 page FR 39133, Federal Register Volume 69, No. 124 June 29, 2004.

www.ingramcontent.com/pod-product-compliance
Lightning Source LLC
Chambersburg PA
CBHW081755170526
45167CB00009B/4028